MINISTRY OF OVERSEAS DEVELOPMENT

OVERSEAS RESEARCH PUBLICATION NO. 17

Plant Diseases of Fiji

by K. M. Graham, B.S.A. (Toronto), Ph.D. (Toronto)
Plant Pathologist

LONDON
HER MAJESTY'S STATIONERY OFFICE
1971

Published by

HER MAJESTY'S STATIONERY OFFICE

To be purchased from
49 High Holborn, London W.C.1
13a Castle Street, Edinburgh EH2 3AR
109 St. Mary Street, Cardiff CF1 1 JW
Brazennose Street, Manchester M60 8AS
50 Fairfax Street, Bristol BS1 3DE
258 Broad Street, Birmingham 1
7 Linenhall Street, Belfast BT2 8AY
or through any bookseller

Price £3. 10s. 0d. [£3. 50] net

SBN 11 880032 9*

Printed in England for Her Majesty's Stationery Office by
Unwin Brothers Limited, Old Woking, Surrey, England.

FOREWORD

Dr. Graham's book will fill a great need in the libraries of extension workers and students in Fiji and much of the Pacific Area.

His study of the plant diseases of the region, and hence the compliation of this book, are the result of a Commonwealth Development and Welfare Grant from 1964 to 1967. During this period Dr. Graham travelled widely both in Fiji and other Pacific territories and a very great deal of this work is based on his own observations. Like any list of its kind it does not pretend to be complete as the area covered in the book is one which has not had the advantage of previous studies in this field but it does represent a complete record of our knowledge to date. I hope it will be the forerunner of similar texts—in fact, a sister volume on the pests of the area is already well advanced in preparation.

The Department of Agriculture is pleased and proud to be associated with this publication and with an author who has come to be a recognised authority on his subject in the Pacific. Unfortunately, Dr. Graham is leaving for fresh fields but the value of his work in the area will be extended long after his departure in this most useful reference.

I would like to express my thanks to the Ministry of Overseas Development and the Commonwealth Mycological Institute for making its early publication possible.

K. J. GARNETT
Director of Agriculture

Overseas Research Publication No. 17
Plant Diseases of Fiji.

CORRECTIONS

Page xvii Figure 186 should read "Corky spot on young papaw fruit".

Page 105 Control, item 2, should read "Keep mildew-susceptible weeds under control".

Page 195 Disease Cycle, line 5, first word should read "unwounded".

MINISTRY OF OVERSEAS DEVELOPMENT

LONDON: HER MAJESTY'S STATIONERY OFFICE

CONTENTS

CONTENTS

CONTENTS

CONTENTS

CONTENTS

CONTENTS

CONTENTS

ILLUSTRATIONS

ILLUSTRATIONS

ILLUSTRATIONS

INTRODUCTION and ACKNOWLEDGMENTS

This manual, a distillation of a series of lectures prepared for third-year students at the Fiji School of Agriculture, makes no claim to infallibility. Such a publication is usually the swan song of a pathologist who has spent many years of residence observing the diseases of a single crop or group of crops. *Plant Diseases of Fiji* is the product of one 3½-year tour as the *vuniwai ni teitei* (doctor of vegetable gardens) at Koronivia. It is hoped, therefore, that the shameless piracy of other people's ideas will be forgiven, and that the "book of words" will not be exceeded by the "book of revision".

The manual was written at the request of the extension division of the Fiji Department of Agriculture. It attempts to describe and illustrate most of the troubles that an agricultural extension worker is likely to encounter in a day's work. Control measures are in most cases based upon the best information available from overseas, because time has not permitted evaluation on the basis of local trials.

Although it was written primarily for application in Fiji, the manual should have wider use. The subsistence and export crops of Fiji, Samoa, Tonga, and other South Pacific islands are very similar. Sugarcane diseases are omitted, because excellent references on this subject are available. Duruka *(Saccharum edule)* diseases are included, because this plant is potentially an economic vegetable crop, and because its cultivation near cane may pose certain problems. Diseases of ornamentals are also omitted, because they have very little economic significance in Fiji at the present time.

It is hoped that the known diseases are effectively separated from the unknown. Future research should attempt to sort out the identity of the virus diseases of both economic crops and weeds. The pathologist and chemist should also combine forces to determine symptoms due to deficiency and excess of trace elements in certain soils, and to investigate the effect of soil fertility on disease incidence. A considerable gap to be filled is the breeding and selection of disease-resistant varieties of rice, cucurbits, tomatoes, and other subsistence crops. This would be the cheapest form of disease control.

The Fiji Islands, though small in extent, have made creditable, if not notorious, contributions to plant disease, with bunchy top in 1891, Fiji disease of sugarcane in 1894, and black leaf streak in 1963. These events do not necessarily indicate a South Pacific garden of Eden where new diseases arise, but the troubles attending the intensive cultivation of two developed crops, bananas and sugarcane. The damage they have caused around the world shows how important it is to recognize early symptoms and to acquire an understanding of the causes and control of disease.

The author wishes to thank the following for their co-operation: Miss S. K. Shankaren, for a very patient and excellent job of typing the manuscript; Mr. Harry Prasad and the darkroom staff of Stinson's Limited, for printing and enlarging services; Mrs. D. M. Keshyap, Miss Marieta Wiliame, and Mr. Vilisoni Fauoro, for technical assistance in field and laboratory work; Mr. P. L. Rhodes and the staff of the Koronivia Research Station, for co-operation in conducting trials on rice, bananas, tomatoes, etc.; the Extension Staff of the Fiji Department of Agriculture, for many specimens of plant diseases, and for assistance on field trips; Lanes Pty. Ltd., ICIANZ (Morris Hedstrom Ltd.), Shell Co. (P. I.) Ltd., and Ivon Watkins-Dow Ltd., for contributions of samples of fungicides and for technical advice; the Commonwealth Mycological Institute, Kew, for assistance in identification of fungous pathogens; Mr. A. W. Allen, Mr. John Parham, and Mr. Gordon Swaine, for preliminary criticism of the manuscript and photographs; Mr. Anthony Johnston, for final criticism of the manuscript and correction of scientific names, and for CMI records of plant diseases in the South Pacific region; to Dr. Dorothy Shaw, Mr. B. Huguenin, and Mr. D. Zaiger, for contribution to distribution records; Mr. John Deering, Agricultural Information Officer, for guiding the manuscript through to publication.

The following illustrations are gratefully acknowledged:
Fig. 15 – Dr. George Zentmyer, University of California, Riverside.
Fig. 35 – Dr. Harry Stover, United Fruit Company, Tela, Honduras.
Fig. 76 – Mr. D. Zaiger, Dept. of Agriculture, U.S. Trust Territory.
Figs. 107 and 108 – Dr. Mario O. San Juan, University of the Philippines, Laguna.
Fig. 109 – Dr. Maria Salome E. Del Rosario, University of the Philippines, Laguna.
Fig. 120 – Mr. Winston Thompson, Fiji Department of Agriculture.
Figs. 124 and 125 – Dr. Dorothy Shaw, Dept. of Agriculture, Stock and Fisheries, Port Moresby, Papua-New Guinea.
Fig. 217 – Plant Diseases Division, DSIR, Auckland, New Zealand.
Fig. 227 (left) – Mr. S. Rumsey, Plant Diseases Division, DSIR, Auckland, New Zealand.

K. M. GRAHAM
Plant Pathologist

Plant Pathology Laboratory
Koronivia Research Station,
Dept. of Agriculture,
FIJI.

THE NATURE AND CAUSES OF DISEASE

Plant diseases rarely destroy an entire crop. Many reduce yield by 5% or less, but the losses are considerable when they are all added up. For example, control of seedborne disease in rice, which is manifested only by spotty stands, gave in one location in Fiji an increase in yield of from 12 bags in the control plot to 20 bags in the treated plot. The profit on an investment of 2 shilling's worth of fugicidal dust was estimated to be about 2400%. In the South Pacific, where land is limited and high yields should be the objective, efficient recognition and control of disease is most important.

A plant disease may be defined in two ways: it is a condition of an agricultural plant which makes it unfit for the economic purpose for which it is intended by man, *or*, it is a condition of any plant, agricultural or wild, which reduces its vigour and changes its appearance so that it looks different from a normal plant.

Wild plants do not usually have serious diseases. Cultivated plants do, because the development of agriculture has upset the balance of nature. Wild plants do not grow in huge populations of single varieties. If they do, they may become devastated by disease. In their natural habitats, they produce only enough fruit, seed, or tubers to survive and carry on the species. Selection and breeding of crop varieties for high yields often leaves the disease resistance of the wild relatives behind. Therefore disease develops when the natural habitat is disturbed, and when a susceptible plant, a disease-causing agent, and a favourable environment combine.

The causes of plant disease are parasitic and non-parasitic. The parasitic causes include fungi, bacteria, viruses, nematodes, algae, and seed plants. The non-parasitic causes include disturbances in nutrition, the wrong environment, bad culture, and the improper use of chemicals. Each of these is briefly discussed:

(1) *Fungi* are plants without chlorophyll or green colouring matter. They may be big enough to be seen without a microscope, like mushrooms or puffballs, or they may be very small. They are all made up of small tubular structures, called *hyphae*, and the mass of hyphae constitutes the *mycelium*. They reproduce and spread themselves about by *spores*, which are very small and easily blown by the wind. Certain fungi, the water moulds, produce swimming spores or *zoospores* that allow them to move about in rain water or wet soil. Spores are produced in or on *fruiting bodies*, that vary considerably in structure and complexity. When conditions are too dry or too hot or too cold for growth, many fungi form thick-walled spores called *chlamydospores*, or resistant bodies called *sclerotia,* formed of tight balls of hyphae.

Because fungi have no chlorophyll, they cannot manufacture their own food from water and carbon dioxide and minerals, like green plants, but they must get it from an outside source, which has the ability to do this. Many fungi live on dead or decaying matter, and they are called *saprophytes.* They take part in the process of decay and they are often highly beneficial as scavengers. The fungi important in plant disease are *parasites.* They attack living plants and they often cause disease when they get out of hand.

A few examples of fungous diseases in Fiji and the South Pacific are black pod and canker of cocoa, Sigatoka disease and black leaf streak of bananas, rice blast, powdery and downy mildews of cucurbits, and rusts of maize and sorghum.

(2) *Bacteria* are very small organisms invisible to the unaided eye because they are about 1/2000th of a millimetre long. They also lack chlorophyll and are saprophytic or parasitic. Some bacteria are *motile*, possessing whip-like swimming organs called *flagellae*, that allow them to swim about in water and infect plants through their stomata or breathing pores.

Bacteria reproduce by simple fission, i.e., they split down the middle. Some bacteria, under favourable conditions, can divide every 20 minutes, grow up, and divide again every 30 minutes, so in 24 hours a single bacterium could produce 300,000,000,000 individuals, if everything were right. In a week's time, the mass of bacteria would equal that of the Earth. Fortunately, competition by other organisms for food supply, and disease among the bacteria themselves, prevents them from getting away with it.

Serious bacterial diseases in Fiji are bacterial wilt of potatoes and tomatoes, and citrus canker.

(3) *Viruses* cannot be seen with the ordinary microscope, and they are considered to be on the borderline between living organisms, e.g., bacteria, and chemical particles. They are known to infect plants and multiply in them largely by the symptoms they produce. They are completely dependent upon plants or animals to reproduce themselves. Usually they cause disease, but symptoms may be mild or severe.

Viruses are known to possess a definite chemical composition. As methods of extracting viruses from their hosts have improved, it is now possible to study their structure. The particles of tobacco mosaic virus, for example, are known to consist of *protein* and *nucleic acid.* The nucleic acid is infectious, and the protein is not. Furthermore, the particles of tobacco mosaic are rod-shaped. Those of cucumber mosaic virus are flexuous rods, like spaghetti. Other plant viruses have spherical particles.

Viruses do not produce spores that swim or blow about like fungi, nor motile cells, like bacteria. Spread of virus disease is accomplished by a transfer of infective plant sap from diseased to healthy plants. This may be done by man, animals, insects, soil, nematodes, fungi, or farm implements and knives. Diseased plants must be injured in some way to release the infective sap. Many viruses are transmitted only by insects, e.g., aphids, leafhoppers, mealy bugs, white flies, and some are transmitted by mites, members of the spider family.

Virus disease in Fiji include tobacco and tomato mosaic, banana bunchy top and mosaic, quick decline and woody gall of citrus, and Fiji disease of duruka and sugarcane.

(4) *Nematodes* are microscopic worms. Although they are animals, like earthworms, they are included in the study of plant disease because their damage resembles the symptoms produced by soilborne fungi and bacteria. Many nematodes are saprophytic, living in decaying vegetable matter, and others are

parasitic. The parasitic nematodes possess a feeding structure called a *spear* or *stylet* whereby they can puncture cell walls and suck out the contents of cells. They may feed inside or outside their hosts.

They attack not only the roots of plants, but also stems, leaves, flowers, grain, or fruit. They can cause stunting, wilting, root rot, galls, and even spots on leaves. Those that produce galls cannot produce root rot, and *vice versa*.

They are spread from plant to plant mainly by movement of soil. Their rate of spread is usually very slow. They can migrate from the soil surface to leaves or flower parts through water on the surface of plants, or by soil blown about by the wind.

Nematodes can form resistant stages, called *cysts,* that are viable for 3-5 years. Seed gall nematodes, that occur in colder climates, have been known to survive for over 39 years.

Nematode diseases in Fiji are root knot of tobacco, tomato, banana, and many other hosts, and banana root rot, caused by the burrowing nematode, *Radopholus similis.*

(5) *Algae* are threadlike or single-celled plants with chlorophyll. Most of them are found in water but some live on the bark of trees in moist places. One or two species are parasitic because they send "roots" or *haustoria* into the epidermal cells of leaves and into bark. They do not normally cause disease unless trees are badly weakened by other causes: shade removal, too much shade, lack of air movement, or low soil fertility.

Like the water moulds, algae are spread by their swimming spores. In Fiji, one species, *Cephaleuros pulvinatus*, is known on coffee, tea, cocoa, avocado, annatto, citrus, guava, and mango, but it rarely causes much damage.

Algae may associate with fungi to produce growths called *lichens.* Somehow the partners get along better together than apart, therefore the combination makes an ideal marriage. The algae produce sugars by photosynthesis, and the fungi use the sugar, and provide a home for the algae. There is some evidence that the fungi fix nitrogen as well. Lichens do not directly cause disease, but they can build up on neglected trees, e.g., citrus, to the point where sunlight is excluded.

(6) *Seed plants* do have parasitic members in a number of families, but they rarely cause disease. These parasites have the same structure as the seed plants that live independently, but they live entirely or partly on other plants, and they may possess or lack chlorophyll. Those possessing chlorophyll are simply using another plant's conductive system for water and mineral supply. Those lacking chlorophyll can germinate their seeds apart from their hosts, and then they become parasitic, sending holdfasts or specialized roots into the bark and wood of their hosts. This group must depend upon its hosts for sugar supply.

Dodder *(Cuscuta campestris)* is a very common plant parasite in Fiji. It has many hosts. In Fijian, it is called *waverevelagi.* If abundant, it can reduce vigour and yield. Another plant parasite is *Cassytha (walutumailagi)* which attacks neglected cocoa coffee, and citrus.

Parasites that do not cause disease include the Fiji mistletoe or *buatoka (Amylotheca insularum),* which decorates forest trees with its showy red flowers, and sandalwood, or *yasi (Santalum yasi),* which starts out in life as a parasite on the roots of other trees, and then becomes independent.

(7) *Disturbances in nutrition* not only cause disease but they influence the degree of severity of parasitic diseases. All plants require a regular supply of nutrients. If there is a deficiency, or an excess, of one or more nutrients, then there is abnormal development.

Plants suffer from deficiencies and excesses of major and minor elements. For example, nitrogen deficiency usually causes yellowing and slow growth, while an excess of nitrogen produces lush growth, lodging, and susceptibility to certain diseases. Rice blast and coconut seedling blight are both favoured by excess nitrogen.

Phosphorus-deficient plants are usually dark green, often developing red and purple colours. Stems are short and root development is poor. The plants are susceptible to root rots and damping-off. Plants with an excess of phosphorus may be more susceptible to certain virus diseases, e.g., tobacco mosaic.

Potassium-deficient plants often show a marginal scorching of leaves and spots of dead tissue between the veins. Stalks are slender and plants are more susceptible to powdery mildew, rusts, and wilts. Cereal grains may be poorly filled. An imbalance between nitrogen and potassium, in favour of nitrogen, may increase susceptibility of tobacco to brown spot, and of tomato and potato to early blight. An excess of potassium may increase suceptibility to organisms that cause overgrowths.

The availability of minor elements is controlled largely by the degree of soil acidity. At a high pH (8-9.5), iron, manganese, copper, zinc, and perhaps boron become unavailable or insoluble by being tied up with other salts. High alkalinity is common on the soapstone soils of the Suva Peninsula, and on any soil with a substratum of old coral. Deficiencies of manganese cause leaf mottling in citrus and tomatoes; of zinc, "little leaf" in citrus, and "white leaf" in maize; of copper, dry fibrous pulp and fruit cracking in citrus; of boron, hollow stem and soft rot in cabbage, cauliflower, and Chinese cabbage.

At a low pH (3.5-5.5), calcium, magnesium, and to some extent, the major elements nitrogen, phosphorus, and potassium, become unavailable. High acidity is common on the red hill soils and also many of the Rewa delta soils. A lack of calcium can weaken a plant's cell walls to the point where invasion by soft-rotting bacteria is very easy. Lack of magnesium causes yellowing similar to that induced by a lack of nitrogen. Diseases such as leaf spot of bananas are definitely most severe on the highly acid soils of Fiji and Samoa. On the somewhat alkaline soils of Tonga (Tongatapu and Ha'apai), bananas are now being grown without fungicidal sprays.

(8) *The wrong environment* often contributes to disease. Agricultural plants are adapted to specific environments. Tropical plants do not survive outside the tropics unless their environment is duplicated in a heated glasshouse. Temperate-

zone crops do not generally do well in the tropics unless they are at a high altitude or grown in cool seasons. Within crops, varieties are also adapted to specific conditions.

If rice, an aquatic plant, is grown under dry land conditions, it suffers badly from "fool's rice" (bakanae disease) because the parasitic fungus thrives in a dry soil. If water-melons, a dry season crop, are grown in a zone with 90-120 inches of annual rainfall, they are destroyed by at least 4 foliage diseases that become extremely difficult to control by spraying. If the dalo variety Samoa, adapted to dry land conditions, is planted in poorly drained soil, it develops corm rot. On the other hand, Samoa stands up better to drought than good wet land varieties. Potatoes, a crop suited to high altitudes and cool nights, does not do well near sea level in the tropics. Yields average 3-5 tons per acre, compared with 12-15 tons normally obtained in temperate climates. The days here are not long enough nor is the soil cool enough to promote maximum tuberization.

(9) *Bad culture* also contributes to plant disease. Farmers know that they must grow most crops in rotation with others. If they do not, soilborne disease builds up and fertility is reduced.

If dalo, peanuts, tomatoes, chillis, eggplant, etc., are grown for too long on the same land, they suffer severely from stem rot, caused by the fungus *Sclerotium rolfsii.* This fungus requires undecayed organic matter to get a start before it invades living stems. If refuse from a previous crop is left untilled or if weeds are not removed from a crop after they are pulled, the fungus has plenty of food. Also, the fungus invades easily from soil piled too high around the stems, or from undecayed stubble and weeds left in close contact.

Damping-off of seedlings is also caused by a number of fungi that multiply in undecayed organic matter. Seedbed and nursery soils prepared from well-aged compost minimize the danger.

Bad culture also includes the neglect of sanitary practices. Pruning dead wood from tree crops goes a long way to controlling canker and dieback. *"Root out and Burn"* is an excellent motto for disease control.

(10) *Improper use of chemicals* includes both the accidental application of fertilizers and weedkillers. Too much or too little fertilizer may result in a serious imbalance of either major or minor elements. Fertilizer applied too close to young seedlings can burn the roots and stems. The injury resembles damping-off.

Certain weedkillers of the hormonal type are active at a very low concentration and they can be applied accidentally in spray machines previously used for killing weeds. They are also extremely volatile and they may blow for miles on a windy day. Symptoms of herbicide injury are curling of leaves, thickening of veins, and fasciation of stems. Often herbicide injury is confused with virus disease.

References: 1, 2, 5, 7, 10, 11, 14, 26, 36, 39, 53, 58, 65, 68, 74, 78, 84.

DAMPING-OFF

Fusarium, Glomerella, Phytophthora, Pythium, Rhizoctonia spp.

Fig. 1. Damped-off tomato seedlings.

Damping-off is a serious problem in seedlings of all crops. If plants become infected in the seedling stage, they may die or survive to be stunted. Yield is reduced or the crops are unfit for market.

Symptoms
Seedlings may rot in the ground before they emerge. This phase is called pre-emergence blight. Gaps in the rows or beds are evidence that this has happened. Emerged seedlings may rot at or just above the soil line, and then they fall over (Fig. 1). This phase is called post-emergence damping-off or blight. If seedlings manage to survive both of these phases, they may develop cankers at the soil line but they remain alive. The canker is usually grey or reddish-brown, and the tissues may be so eroded that the stems are constricted. This condition is often called wire-stem. Plants with these basal cankers are unthrifty.

Disease Cycle
A number of soilborne fungi can cause damping-off. They have one thing in common: they are associated with large amounts of undecayed organic matter. These fungi attack seedlings usually in the region of the hypocotyl, or the transition zone between the root and the stem. If plants get through the emerging phase, certain fungi, e.g. *Rhizoctonia,* can still attack and erode the epidermis and underlying cells.

Other conditions that favour damping-off are lack of sun, too much moisture, and too much nitrogen in the seedbeds.

Control

1. Make sure that the seedbed is relatively free of undecayed organic matter. Do not get soil from a source that has just grown a crop.
2. Drench the seedbed with formaldehyde (1 gallon of 40% formalin concentrate in 50 gallons of water) at the rate of 2 pt/sq ft. Cover the bed with a tarpaulin or sacks for 24 hours. Do not sow seed for 10 days.
3. Do not use stagnant water for watering seedlings.
4. If damping-off appears in the seedbed, drench the surface thoroughly with captan (1 oz/3 gallons).

References: 14, 36.

DODDER INFESTATION

Cuscuta campestris

Dodder is very common in Fiji and probably elsewhere. It is a frequent parasite of roadside weeds in Viti Levu, Vanua Levu, and Taveuni. It has not been seen in cultivated crops, except citrus. Its economic importance is not known.

Fig. 2. Reed *(Miscanthus)* infested by dodder.

Symptoms

Infested hosts are covered by leafless networks of yellowish wire, about ⅛ inch in diameter, or less (Fig. 2). Buttonlike flowers and seed capsules are produced on these growths. Severely infested plants do not yield as well as they should.

Disease Cycle

Seeds of dodder germinate in the soil. The seedlings attach themselves to seeds and low-hanging twigs of trees by holdfasts *(haustoria).* These growths penetrate stems and align themselves with the vascular system of the host plant, and suck out nutrients coming up from the soil. Then they lose their soil connexion and become completely dependent upon the host plant. Birds, animals, and man may carry the seed to uninfested areas.

The host range of dodder is unlimited. In Fiji it has been found on many weeds and grasses. It probably was imported accidentally in seed from elsewhere.

Control

1. Cut and burn all infested plant parts.
2. Spray the area thoroughly with the appropriate herbicide and keep it in a state of fallow for one year.

References: 7, 62.

Improper Use of Chemicals

Fig. 3. 2,4-D damage to cowpea.

Fig. 4. 2,4-D damage to dadap.

Fig. 5. 2,4-D damage to citrus.

Fig. 6. Dalapon damage to banana.

Fig. 7. Dalapon damage to papaw.

Fig. 8. 2,4-D damage to kumala.

The careless use of herbicides can cause considerable damage to all kinds of crops. It is important to recognize the injury when it occurs.

Symptoms and Cause

The hormone-type herbicides, e.g., 2,4-D, 2,4,5-T, MCPA, etc., cause twisting, cupping, vein thickening, and stunting of the foliage of many crops and trees (Figs 3, 4, 5, 6, 7, 8). Some crops, such as tobacco, tomato, papaw, cucurbits, and legumes, are very sensitive to very small quantities of herbicide. If the chemical is applied to roadside weeds on a windy day, traces of it may drift for miles and affect neighbouring crops. The most common source of contamination is a knapsack sprayer or misting machine used first for the application of herbicide and then for the application of a fungicide or insecticide.

Control

1. Apply herbicides according to the directions of the manufacturer and the Weed Control Officer.
2. Use separate machines for the application of herbicides. Do not use them for any other purpose.

References: 58.

Sclerotium rolfsii

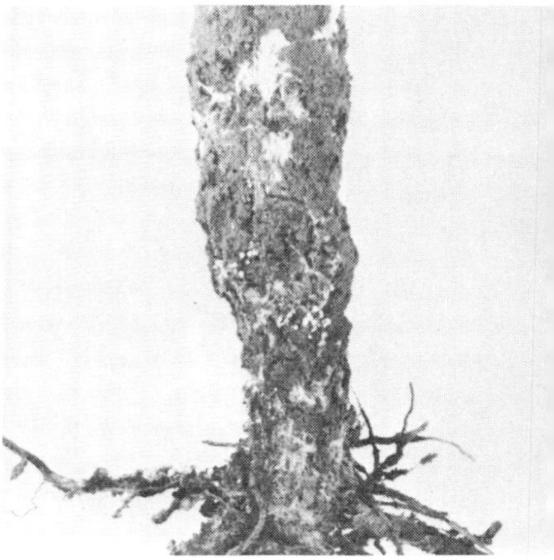

Fig. 9. Stem rot of peanuts (above left).

Fig. 10. Stem rot of chilli with pinhead-sized white sclerotia of *Sclerotium rolfsii* (above right).

Fig. 11. Mycelium and sclerotia of *S. rolfsii* growing over soil at the base of a potato stem (left).

Stem rot is a serious disease of all vegetables. It may also attack cereals and grasses, and even tree crops in the seedling stages. It is much worse on some soils than on others.

Symptoms

Plants wilt or they drop their leaves. The bases of the stems show a white fungus growth with pinhead-sized white lumps (Figs. 10, 11), that grow to about the size of a radish seed. They turn a tan colour when they are fully developed. Stem tissues underneath the growth are shredded, blackened, and eroded in appearance (Figs. 9, 10). Roots may also be rotted.

Disease Cycle

The stem rot fungus can be found in undecayed organic matter from any kind of crop. The disease is worst when insufficient time is allowed for the residue from a previous crop to rot down. If the soil is poor, e.g., talasiga soil, the rotting process is very slow and *Sclerotium rolfsii* builds up. Weeds that have been pulled and left to die in the rows may also provide food for the fungus. If soil is heaped up too high around stems, and if it contains undecayed trash, the stems will be attacked.

Control

1. Rotate highly susceptible vegetable crops, e.g., chillis, tomatoes, potatoes, with rice, maize, sorghum, and grasses, that are relatively resistant.
2. Plough under crop residues to a depth of 9-10 inches and top dress with a high-nitrogen fertilizer to stimulate the rotting action of soil micro-organisms.
3. Do not hill up too high or throw weeds into the rows. Plant peanuts on ridges. Hill up potatoes only after the tubers have started to form.
4. Pull and burn diseased plants as soon as they are detected.
5. On high-value annual crops, e.g., peanuts, experiment with PCNB (quintozene, Pura-soil, Terrachlor). Dust on the ridges at pegging time at the rate of 15-20 lb per acre.

References: 3, 5, 14.

ANTHRACNOSE and FRUIT ROT

Botryodiplodia theobromae, Glomerella cingulata

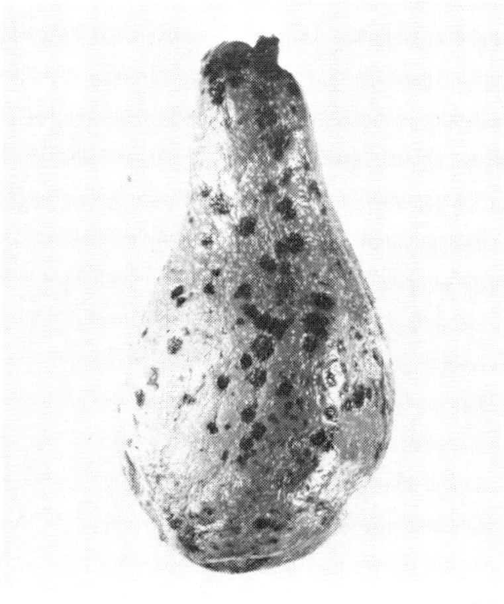

Fig. 12. Anthracnose of avocado.

Fig. 13. Botryodiplodia fruit rot of avocado.

Anthracnose and ripe fruit rot are unavoidable in perishable fruit such as avocadoes. It is not important locally unless fruit is exported to New Zealand.

Symptoms
Anthracnose commences as black round spots about the size of pinpoints. They enlarge to form sunken, circular, dark brown spots about ¼ inch in diameter (Fig. 12). These may cover the entire fruit. The surface may crack and fungus spore masses are exuded in colours of white or pink depending upon the species involved.

Disease Cycle
The anthracnose fungus, *Glomerella cingulata,* establishes latent infections in or near lenticels on the surface of green fruit. It may be followed by a secondary fungus, *Botryodiplodia theobromae,* which may also enter through insect wounds, mechanical injuries, or the break at the stem end (Fig. 13). Sunburn on thin-skinned varieties growing in the wrong environment may also be a starting point. These fungi are always present on dead twigs and fruit.

Wet weather at or near harvest favours anthracnose.

Control
1. Prune dead twigs and branches.
2. Avoid bruising fruit and harvest early.
3. Do not plant thin-skinned varieties in areas with intense sunlight.
4. Spray fruit within 2 weeks of harvest with maneb (½ lb/2 gal) by misting machine. Add 1 tbsp of X-77 or Agral LN to the water before mixing in the fungicide.

References: 90.

DIEBACK and ROOT ROT

Phytophthora cinnamomi

Fig. 14. Avocado tree in an advanced stage of dieback.

Fig. 15. *Left:* healthy root system of an avocado seedling; *right:* two seedlings affected by root rot.

Dieback and root rot is the most destructive disease of avocadoes in Fiji.

Symptoms

Foliage becomes sparse, leaves turn yellowish-green and droop. The new leaves are smaller than normal. Dieback commences at the top of the tree (Fig. 14). For a while, fruit set may be higher than normal, but fruit size is reduced. After the roots have completely rotted, the tree dies.

Roots show blackened sections and the very fine feeder roots are killed (Fig. 15).

Disease Cycle

The pathogen, a water mould, has a very wide host range and it is also responsible for root rot of passion fruit, pineapple, papaw, and dalo. It is a soilborne fungus that is capable of living for long periods as resistant *oospores* and *chlamydospores.* In wet soils the fungus produces *zoospores* that swim about in water. Roots are attacked when air is cut off from them by poor drainage.

Control

1. Plant avocadoes only in well-drained sites.
2. Avoid the introduction of the disease into seedbeds by fumigating them with methyl bromide, and treating the seed with hot water for 30 minutes at 120°F.

References: 90.

ANTHRACNOSE

Colletotrichum musae

Anthracnose or ripe fruit rot is a transport and ripening problem. It can be found everywhere on ripening bananas. Up to a certain stage, it is merely an indication of the degree of ripeness, and very little can be done about it. It has very little effect on internal fruit quality.

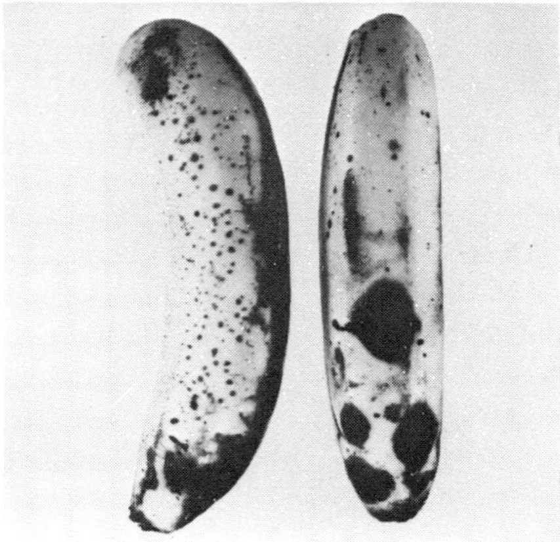

Fig. 16. Two stages of anthracnose development on Veimama.

Symptoms
Small, round, slightly sunken spots appear all over the surface of ripening fruit (Fig. 16). They eventually enlarge to cover the whole fruit. Often anthracnose and finger rot develop at the same time. In a moist atmosphere, the surface of the spots becomes covered with pink spore masses.

Disease Cycle
Immature fruit has latent infection that commences very early in the growth period. The anthracnose fungus penetrates the epidermis but it does not develop further until the fruit starts to ripen. The source of spores is probably old decaying flower parts and leaves. The spores are produced in great masses during rainy periods, and they are splashed from trash on to fruit surfaces. Some infection may also occur in dirty packing sheds and under banana leaves used to cover bunches before they are packed.

Anthracnose may be more serious during the wet season when banana skins are softer, and conditions are more favourable for the fungus.

Control
1. Keep plants free of dead leaves by frequent detrashing. Do not cover unpacked bananas with dead leaves.
2. Pack fruit and get it into ship holds as soon as possible after harvesting.
3. Handle fruit as carefully as possible. Every injury makes anthracnose worse.
4. If bananas are to be washed, use running water from a clean source.

References: 65, 73, 85.

BLACK CROSS SPOT

Phyllachora musicola

This disease is more a curiosity than a problem. It is more common on the ABB bananas (Vudi, Paka, Bluggoe) than on Robusta types. It has been observed in Fiji, Samoa, New Guinea, New Hebrides, and as far north as Taiwan and the Philippines.

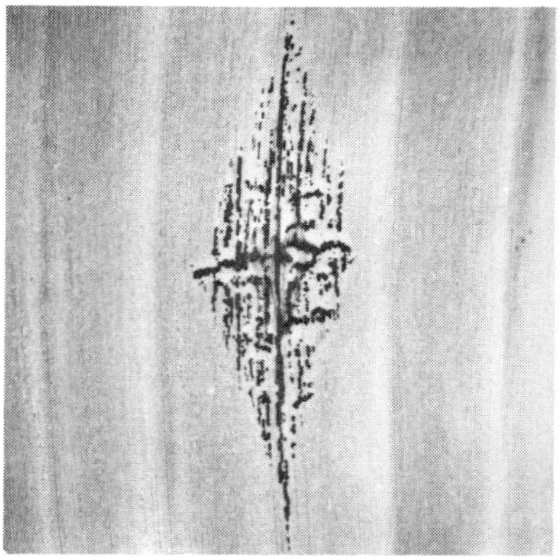

Fig. 17. Black cross spot on Fijian "Vudi" banana.

Symptoms

Spots appear as black streaks with short transverse lines. They enlarge to form an intricate pattern of parallel and transverse black lines with an elongate diamond outline (Fig. 17). The spots look as though they had been etched into the leaf surface. They are most numerous towards the edges of the leaves. Under very favourable conditions, the lower leaves may be killed.

Disease Cycle

The fungus produces *ascospores* in flasklike structures *(perithecia)* embedded in the leaf tissue. The life cycle is probably similar to that of *Mycosphaerella musicola,* the cause of Sigatoka disease.

Control

Black cross disease is so rare on commercial bananas that no special control measures are recommended.

References: 30, 56, 70, 85.

BLACK SPECK

Phosphorus deficiency

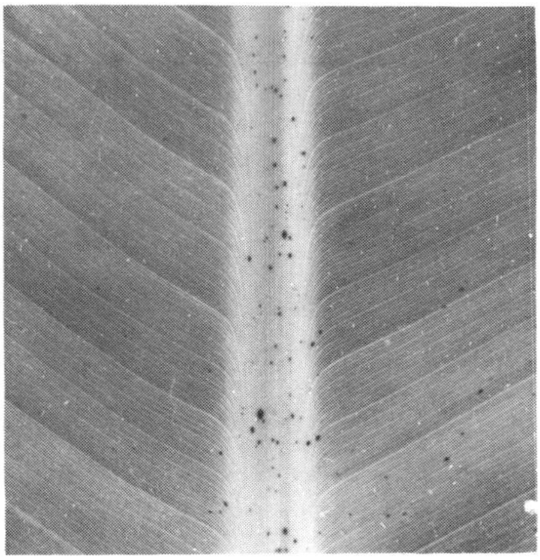

Fig. 18. Specks on midrib.

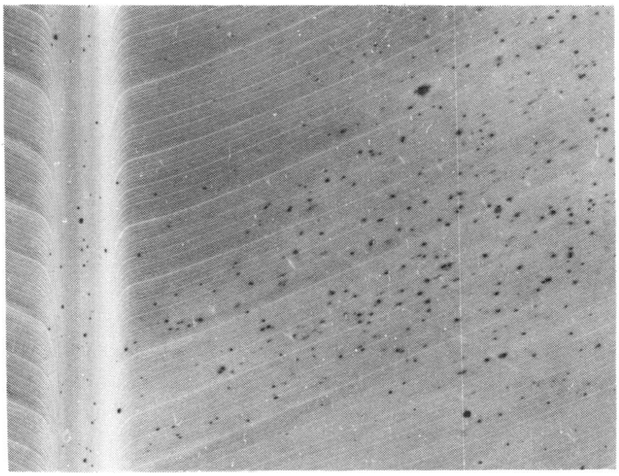

Fig. 19. Specks and streaks on lamina.

This trouble is common in bananas growing on hill soils low in fertility. It can also occur in plants that have not been fertilized regularly.

Symptoms

Black spots of pinhead size appear on the midribs (Fig. 18). They may also appear on the laminae or they may be mixed with short black streaks (Fig. 19). Similar spots caused by *Cercospora* are reddish-brown and much larger or longer.

Plants are stunted, yellowed, and leaves tend to choke at the throat. Leaves eventually scorch and die.

Cause

Pot experiments have shown that black spots are associated with a phosphorus deficiency. It is common where phosphorus is not available in highly acid soils.

Control

1. Apply a high-phosphate fertilizer according to the directions of the Soil Chemist.
2. Adjust soil pH by the application of lime.

References: 85.

BLACK SPOT

Sucking Bugs

Black spotting of fruit has been known to occur for several years from certain areas of Fiji. It has never been serious.

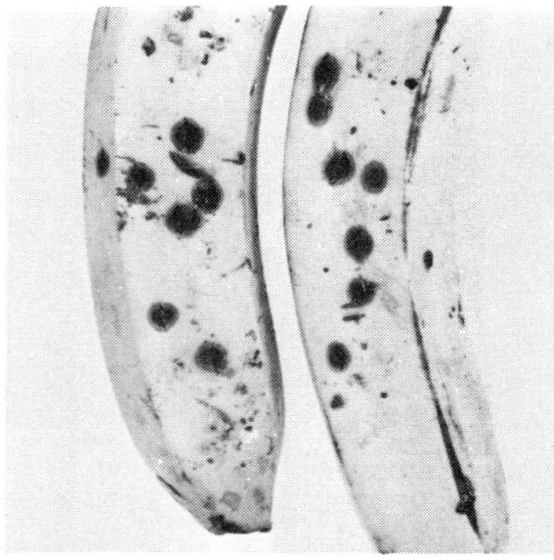

Fig. 20. Black spotting of Veimama.

Symptoms
Black spots about ¼ inch in diameter occur on any part of the fruit. They look like small craters (Fig. 20). Sometimes they are overgrown by a reddish-brown mould. The edges of the spots are watersoaked in appearance.

Cause
Black spot is generally attributed to the feeding of sucking bugs. The trouble appears usually towards the end of the cool season. It has been common in bananas shipped from the south coast of Viti Levu. Secondary fungi may invade the spots but the fruits are apparently attacked when they are young enough to put up a strong defense reaction by laying down cork barriers.

Control
No special measures are recommended.

References: 65.

Bunchy Top Virus

Fig. 21. "Dots and dashes".

Fig. 22. Healthy leaf.

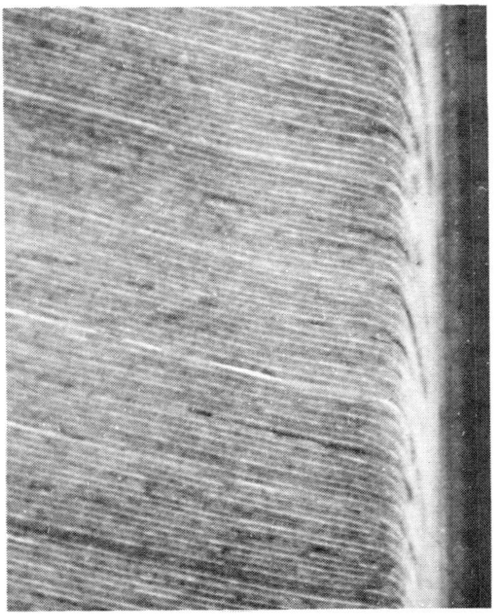

Fig. 23. "Hooks" along midrib.

Fig. 24. Severe vein clearing.

Fig. 25. Infected sucker with narrow, bunched leaves. (above left)

Fig. 26. Infected sucker with scorched leaves. (above right)

Fig. 27. Streaks on midrib. (left)

Bunchy top is very serious in the banana-producing islands of the South Pacific. In Fiji it caused a serious reduction in exports during 1892-95, 1900-1905, and in Samoa it probably reduced exports by 30-40% during 1963-1966. It is also serious in Tonga. The disease has also been reported from the Ellice Islands, Wallis Islands, and New Caledonia. It has not been found in the Cook Islands nor in Papua-New Guinea.

Symptoms

If the base of the 2nd or 3rd leaf of a suspected plant is held up to sunlight and examined from the underside, narrow dark-green lines broken in dot-dash pattern (Fig. 21) are seen lying between the veins and parallel to them. They continue into the midrib of the leaf as dark-green "hooks" (Fig. 23) across the clear areas on both sides of the midrib. Dark-green streaks are often seen on the back of the petiole (Fig. 27). Sometimes a vein clearing (Fig. 24) is also seen as the first symptom of infection. During very cool weather, vein clearing is more obvious.

In the more advanced stages the younger leaves are erect, shorter and narrower than normal (Fig. 25), and yellowed along the edges. The yellowing often turns into a marginal scorch (Fig. 26). The leaves are brittle and if they are broken they snap off crisply. Plants are stunted and the leaves are choked at the throat. Roots may be rotted. If a bunch is formed, it is small and useless.

Sometimes the only sign of infection in a plant is a single yellow unfurling leaf. This symptom can be confused with nitrogen deficiency but the pattern of bunchy-top distribution is much spottier in a plantation.

Disease Cycle

Bunchy top is spread by planting material and by the banana aphid, *Pentalonia nigronervosa.* It *cannot* be spread by knife or juice. Other hosts are wild bananas and species of *Heliconia.*

Suckers often show no symptoms at the time of planting and later they develop them. This is called *latent* or masked infection. Suckers may be infected very early after they emerge by aphids that prefer very young leaves. Sucker-to-sucker transmission is probably very common. Aphids feeding deep in the unfurling leaf of an infected plant may also pick up the virus. They may then leave the plant or sucker by flying away or by being carried away by plantation workers. Aphids are very light and they can be blown for long distances.

When infectious aphids feed on healty plants, they transmit the virus to them. In a single mat, the virus travels down the sap stream from the point of aphid feeding into the corm and out into the young suckers. These are the starting points for a fresh outbreak of the disease.

Control

1. Select original planting material very carefully. If a nursery is to be established, spray at weekly intervals with Metasystox or Anthio (1 oz/10 gallons water), paying particular attention to very young suckers.
2. Inspect nurseries and mature plantations regularly, using the "hook" system of bunchy-top identification.
3. Spray detected plants with Metasystox or Anthio *before* taking them out, paying close attention to all the suckers.
4. Dig out the plants, making sure that no buds are left to start the disease off again. Dig out all the plants in a mat and cut them into small bits.
5. If a plantation is more than 50% infected, destroy the lot rather than try to keep up with replants.

References: 65, 74, 85.

CORDANA LEAF SPOT

Cordana musae

Cordana or large diamond leaf spot is much more serious in Samoa than in Fiji or Tonga, probably because there are more wild bananas *(Musa banksii)* in Samoa. It is a disease that develops in shade.

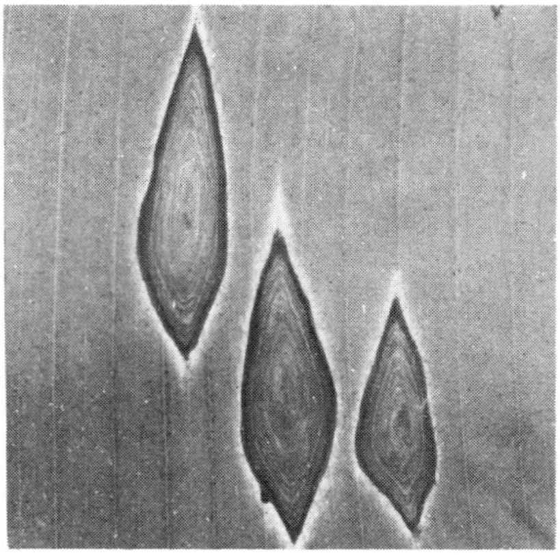

Fig. 28. Cordana leaf spot on the Veimama variety.

Symptoms

The spots begin as egg-shaped black dots, and when they are fully developed, they are large, grey-brown in colour, and roughly diamond-shaped (Fig. 28). They often show a zonate pattern and they may form around a spot caused by Sigatoka disease, black leaf streak, or Malayan leaf spot. They are surrounded by a distinct bright yellow band. The spots may be 3-4 inches long and 1 inch wide. In severe attacks, the whole leaf dies.

Disease Cycle

It is believed that *Cordana* does best with some other fungus such as the pathogens of Sigatoka disease and black leaf streak, or of Malayan leaf spot, to initiate infection. It may then grow out and over into healthy leaf tissue. The spores of *Cordana* are formed on long brown stalks *(conidiophores)* and they can be seen easily with the aid of a hand lens.

Best conditions for disease development are believed to be shade, cool weather, and the presence of other leaf-spotting diseases.

Control

1. Do not grow bananas in deep shade, and do not plant suckers too closely.
2. Use the winter schedule of maneb-oil-water emulsion recommended for control of Malayan leaf spot.

References: 65, 85.

FINGER ROT

Botryodiplodia theobromae
Ceratocystis paradoxa
Colletotrichum musae

Finger rot is a major cause of deterioration of bananas during shipment from the islands and after landing in New Zealand.

Symptoms

As the fruit ripens a brown to black rot commences at the broken or cut ends of the fingers (Fig. 29) and it extends into the fruit. On hands, the rot commences on the cushions and progresses down each finger.

Fig. 29. Finger rot of Veimama.

Disease Cycle

The fungi associated with finger rot normally live on decaying plant parts. They cannot invade unwounded fruit. Any harvesting or shipping operation that injures fruit makes them susceptible to invasion by these wound parasites. Spores of the fungi are present on the fruit in the field. When a wound develops or bananas start to ripen, the spores germinate and invade.

Control

1. Do not harvest fruit for packing that is too ripe.
2. Handle bunches carefully. For the hand trade cut the hands off with part of the cushions intact instead of twisting them off.
3. Paint cushions with Basofix if hands are to be shipped. If fingers are preferred, wash them in a solution of Shirlan or Musal W (4 oz/15 gallons) for 2 minutes, and then rinse with water before packing.
4. Avoid unnecessary delay in getting the fruit from the field into ships' holds.

References: 73, 85.

FRECKLE

Macrophoma musae

Freckle is not important on commercial bananas in Fiji and Tonga, but it is serious in Samoa. In Fiji it is more common on the plantain or "Vudi" types.

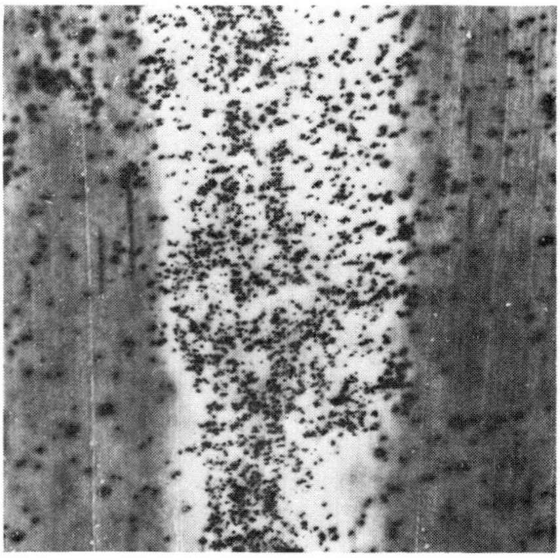

Fig. 30. Freckle on "Vudi" banana leaf.

Symptoms

The spots on the leaves are pin-points (Fig. 30) resembling those of speckle, but they are raised and they feel like sand-paper. They are most abundant on the upper surface. On the fruit the entire surface may be covered by black dots, each of which has a narrow, watersoaked margin.

Disease Cycle

Conidia are produced inside the round, black pinpoints that are the fruiting bodies *(pycnidia)* of the fungus. During rainy weather, the conidia are pushed out and they are washed off the leaves on to the fruit. Then secondary infections develop.

The disease appears to be more serious towards the end of the wet season than during the dry season.

Control

1. Cut off the most severely affected leaves to protect the fruit from infection.
2. Spray with maneb-oil-water emulsion as for control of Malayan leaf spot and Cordana.

References: 56, 85.

MALAYAN LEAF SPOT

Haplobasidion musae

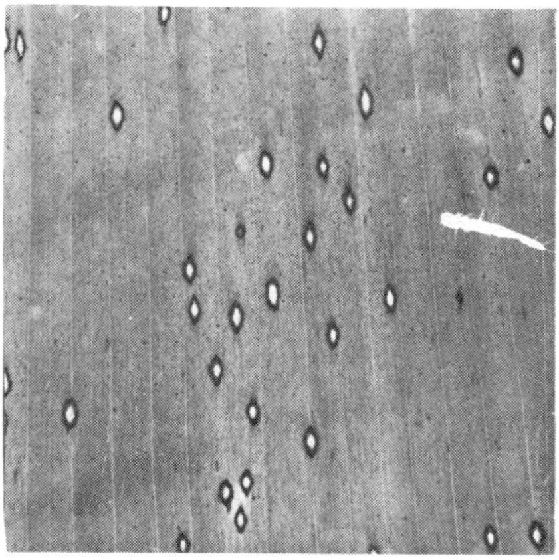

Fig. 31. Malayan leaf spot on the variety Veimama.

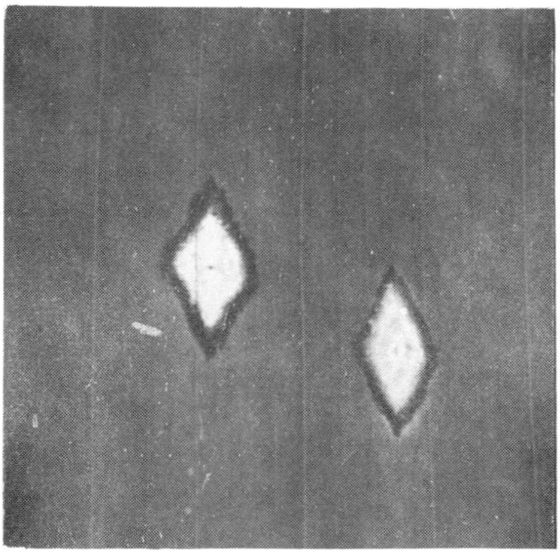

Fig. 32. Typical diamond-shaped spots on older leaf of Veimama.

Malayan leaf spot has been observed in Fiji, Tonga, Samoa and a number of other South Pacific islands. It is more obvious during the cool season, particularly on oil-sprayed bananas. During a severe attack, it may destroy 20% of the leaf surface.

Symptoms

Black dots appear that enlarge to form diamond-shaped spots about ¼-inch across (Figs. 31, 32). The centres turn white, and they may be covered on the undersides with a thick dust of reddish-brown spores. Eventually, leaves are killed, but not nearly as rapidly as with Sigatoka disease and black leaf streak.

Disease Cycle

The life cycle has not been studied in detail. The spores are produced in cushion-like clusters, and they are probably wind and waterborne. The disease appears to be worse in shaded locations with very high rainfall, and it tends to fade out in the warm season. For some reason, probably because of a darkening and shading of the leaf, bananas sprayed with misting oil without fungicide suffer severe attacks.

Control

1. In areas where oil is in use for the first time, observe the severity of Malayan leaf spot throughout the year. It may not be serious enough to justify the use of a fungicide.
2. Where it is consistently serious, spray during the cool season, from May to October, with maneb-oil-water emulsion, at the rate of 2 gallons per acre per fortnight. Use 1½ lbs maneb, 1½ gallons misting oil, ½ gallon of water, and 75 c.c. Triton emulsifier (Kemonic) added *first* to the oil before mixing in the water and fungicide.

References: 30, 56.

MOSAIC

Cucumber Mosaic Virus

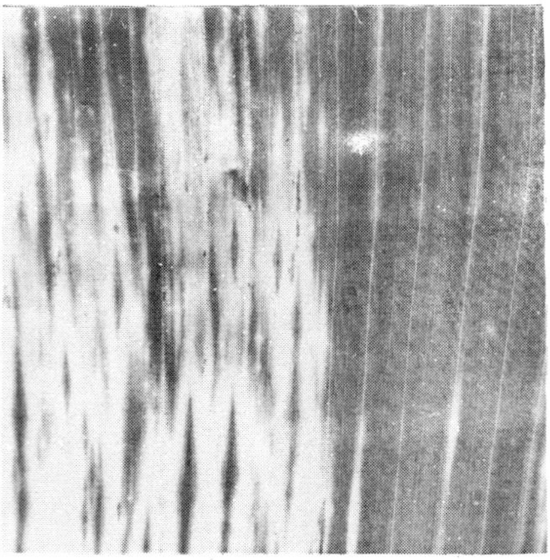

Fig. 33. Chlorotic streaks of mosaic.

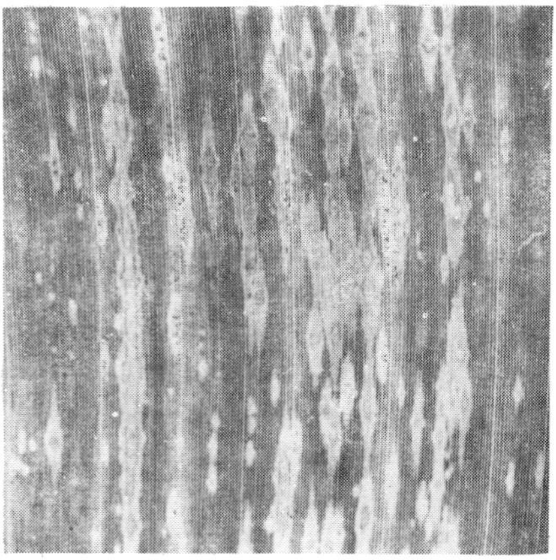

Fig. 34. Chlorotic streaks beginning to turn black.

Mosaic was first seen in Fiji at Koronivia in 1965. It has also been observed in Samoa and Tonga. In the past, it has probably been confused with bunchy top, which it resembles in some respects. The Lady Finger variety is suspected of being a tolerant carrier of mosaic.

Symptoms

Light green or yellow streaks and bands form on the younger leaves (Fig. 33). The unfurling leaf may be almost white, and it may develop a soft, black rot that extends right down into the corm. This heart rot phase is much more likely to develop during the cool season than in warm weather. The streaks on mature leaves may die out to produce a black streaking (Fig. 34) or shot-hole effect.

Mosaic can be distinguished from bunchy top by the absence of the dark green "hooks" along the midrib and "dots and dashes" on the laminae, and the absence of bunching and stunting of the plants.

The effect of mosaic infection on shape and yield of fruit is not clear.

Disease Cycle

The virus that causes banana mosaic and heart rot has a very wide host range. Cucumber mosaic has been observed on cowpeas, chillis, squash, cucumbers, passion fruit, wild cape gooseberry *(Physalis angulata)*, goatweed *(Ageratum conyzoides)*, nightshade *(Solanum nigrum)*. Maize and other members of the grass family may become infected. All of these plants are interplanted or they may be weeds in bananas.

The melon aphid *(Aphis gossypii)*, and the maize aphid *(Aphis maidis)* are suspected to transmit cucumber mosaic from interplanted crops and weeds to banana. Both of these species can feed on a wide range of plants. The banana aphid *(Pentalonia nigronervosa)* cannot, as far as is known, transmit mosaic.

Mosaic outbreaks are associated with excessive intercropping of bananas with susceptible plants such as cowpeas, squash, and cucumbers, etc., and the occurrence of large amounts of the weeds mentioned.

Control

1. Avoid cowpeas, beans, squash, cucumbers, and maize as intercrops with bananas. Keep weeds under control.
2. In the case of an outbreak, use the same control measures as for bunchy top.

References: 65, 74, 85.

PANAMA DISEASE

Fusarium oxysporum f. sp. *cubense*

Some pathologists have reported Panama disease in the South Pacific, and others say that these reports have never been properly confirmed by isolation of the pathogen and inoculation into healthy, susceptible plants. Apparently the disease has been found everywhere else in the world on the Gros Michel variety. Gros Michel has not been grown commercially in Fiji since 1920 and it has never been important in Samoa and Tonga. The popular Veimama, Cavendish, Vudi varieties are resistant to the common race of the pathogen. It has caused serious losses wherever it has been found.

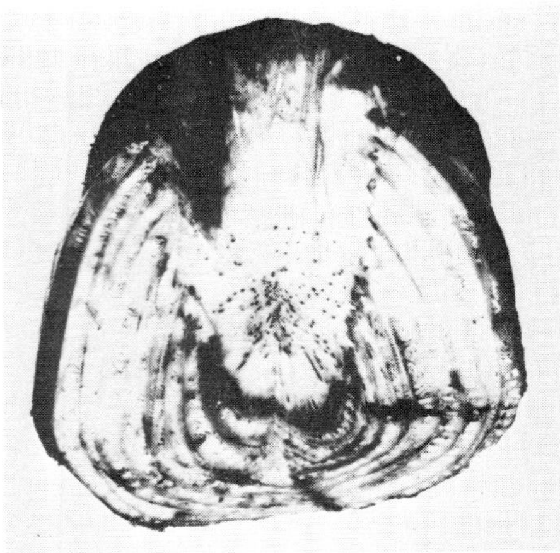

Fig. 35. Dark, discoloured vascular strands in section of banana rhizome and pseudostem base affected by Panama disease.

Symptoms

The margins of the lower leaves develop a deep yellow colour, and then they turn brown and dry out. The petioles collapse around the pseudostem. Sometimes the pseudostem splits up from the ground level.

The most distinctive symptoms are internal. If the base of the plant is cut through vertically, numerous brown and black lines can be seen running in all directions through the corm and upwards along the vascular tissue into the leaf bases and petioles (Fig. 35).

Land infested by Panama disease must be taken out of production for several years.

Disease Cycle

The pathogen is a soilborne fungus which is capable of living in the soil for years in the absence of bananas. When a susceptible variety such as Gros Michel is planted in infested land, the fine lateral roots become infected. The fungus grows up through the vascular system into the corms, where it blocks the movement of sap and causes wilt. The fungus can live between crops of bananas in the roots of a number of species of grasses, and also on banana relatives such as *Heliconia.* It can also form thick-walled *chlamydospores* that survive for years in dry soil.

Panama disease is most serious in acid alluvial soils.

Control

1. Grow resistant varieties such as Veimama and Cavendish.

References: 56, 70, 75, 85.

ROOT KNOT

Meloidogyne incognita

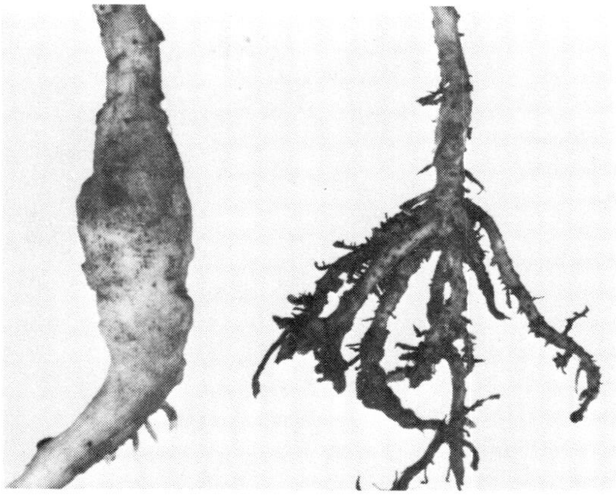

Fig. 36. *Left:* Large banana root with spindle-shaped knot; *right:* small root almost destroyed by rootknot nematode.

Root knot is often found on banana roots along with root rot, caused by the burrowing nematode. It is not as important as root rot.

Symptoms
Spindle-shaped swellings develop like beads on a string along the smaller roots (Fig. 36). Sometimes they can attain the size of a fingertip. Often lateral roots develop around a knot. If the knots are numerous, root rotting may occur.

Disease Cycle
The root knot nematode occurs on many plants, including ginger, dadap, tomato, tobacco, haricot bean, chilli, carrot, sugarcane, coffee, and many weeds. The life history and means of transmission of the nematode on banana are the same as that described for carrot.

Control
1. Use the same measures as for control of burrowing nematode.

References: 17, 18, 33, 36, 85.

ROOT ROT AND "FALL DOWN"

Radopholus similis

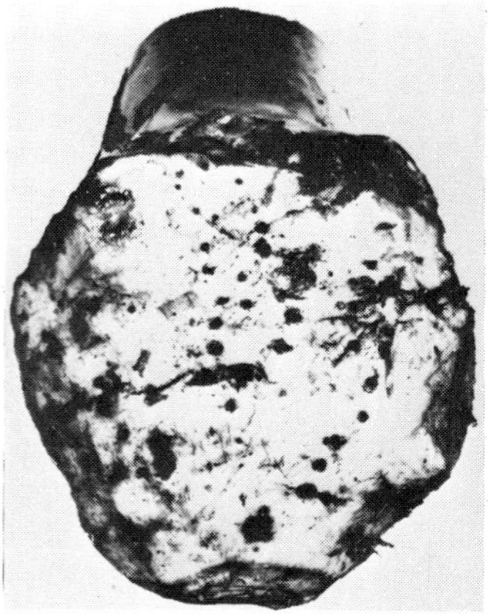

Fig. 37. "Fall-down" caused by root rot (above left).

Fig. 38. "Blackheads" in corm (above right).

Fig. 39. Root rot following nematodes (left).

Root rot has been known in Fiji since 1890 and it is a serious problem on bananas everywhere in the South Pacific. It is mainly responsible for declining yields after 3-4 years in one location.

Symptoms

Young plants turn yellow and stop growing, as if they were suffering from drought. Older plants must be propped after the bunch is thrown, or they will fall down (Fig. 37) in a stiff breeze. Roots are rotted (Fig. 39) and sometimes the rot extends into the corm, producing "blackheads" (Fig. 38). Plants may also look "choked" as if they were suffering from bunchy top.

Disease Cycle

Burrowing nematodes can enter banana roots at any point. They break down the cell walls and feed on the contents. As they move through the tissues, the females lay numerous eggs and colonies of nemas develop along their paths. These tissues are then invaded by secondary fungi and bacteria that produce rot. The nematodes themselves are always found in healthy tissue ahead of the rot. When roots are completely rotted, the nematodes leave in search of a new home.

The burrowing nematode has a very wide host range apart from bananas. It includes tea, coffee, sugarcane, citrus, kumala, bean, tomato, maize, cucurbits, ginger, and *Heliconia.* It has not been reported on dalo, tapioca, and para grass. There may be strains of *Radopholus similis* specialized on certain crops but not on others, e.g., the strains on citrus and sugarcane may not be able to attack bananas and *vice versa.*

Control

1. Select planting material with extreme care.
2. Trim corms of all "blackheads" that are indications of the extent of nematode invasion.
3. Dip corms in hot water held at 124°–126°F (51°–52°C.) for 20 minutes

   OR

   Dip corms in a solution of Fumazone (Nemagon) (40 oz/100 gallons of water) for 1 minute.
4. Do not grow bananas for more than 3 years on the same land. Plant a new area at the end of the 2nd year. Leave infested areas out of bananas for at least one year.
5. Experiment with injections of Fumazone (Nemagon) every 6 months. Inject at 6" depth at 14 points per mat with 10% dilution during dry weather.

References: 17, 33, 36, 79, 85.

RUST

Uredo musae

Rust builds up on Lady's Finger, Dwarf Cavendish, and Veimama during the cool season in the South Pacific, but it has never caused defoliation.

Fig. 40. Rust on Veimama.

Symptoms

Rust can be confused with Sigatoka disease or black leaf streak, but the brown streaks (Fig. 40) are raised on the lower surfaces of the leaves, and they are rough to the touch. The spots are about ¼ inch long.

Disease Cycle

The disease is spread by *uredospores* produced in the pustules *(uredia)* on the lower surfaces of the leaves. It is not known whether young leaves only are infected, as is the case with leaf spots caused by *Cercospora.*

Cool weather and some shade appear to be most favourable for rust.

Control

No special measures are recommended.

References: 56, 85.

Mycosphaerella musicola, Mycosphaerella sp.

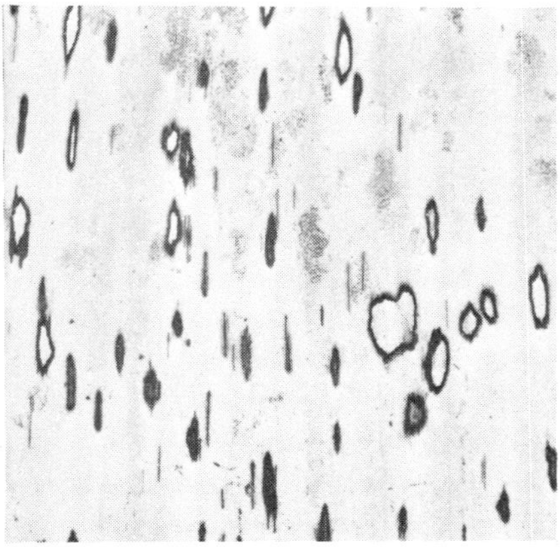

Fig. 41. Sigatoka disease.

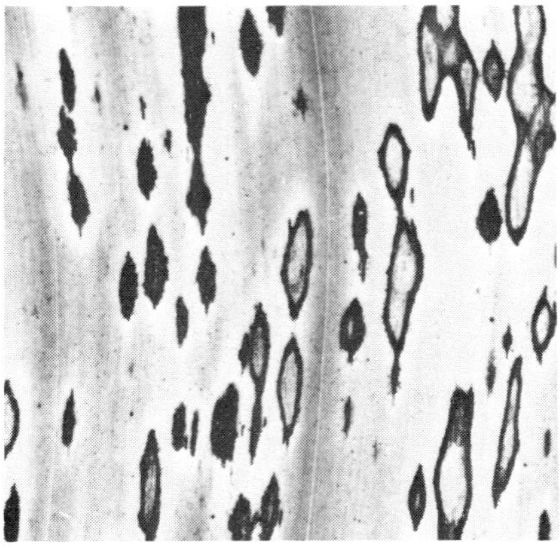

Fig. 42. Black leaf streak.

Sigatoka disease and black leaf streak are regarded by some pathologists as distinct fungal diseases, and by others, as caused by two different strains of the same fungus. They are found everywhere in the South Pacific on bananas with the exception of the Cook Islands. They seriously limit production in Fiji, Samoa, and parts of Tonga, probably by 50-60%.

Symptoms

Spots appear on the leaves first as yellow streaks $^1/_8-^1/_3$ inch long, parallel to the veins. This stage is not seen in black leaf streak during the warm season. The streaks enlarge to form brown or black streaks which dry out to dirty white spots with black pinpoints covering them (Figs. 41, 42). These are the *sporodochia* or spore cushions of the Cercospora or conidial state of the fungus. In a severe attack, particularly by black leaf streak, the spots run together and the whole leaf may die, beginning at the tip and the edges.

Other effects of a severe attack are undersized bunches, short fruit, uneven ripening, and cracking of fruit.

Disease Cycle

Two types of infection occur: (1) The *conidia* produced on the sporodochia are washed off the leaves by rain or dew. They are responsible for spread of the disease over short distances. They are splashed on to the young unfurling leaves. About one month later, the first spots appear, and about 2 weeks later, new conidia are produced; (2) *Ascospores* are ejected from *asci* in *perithecia* or flask-shaped structures that form underneath the sporodochia in dead leaves. The ascospores are very small and light. They are spread by wind over long distances. Like the conidia, they can infect only the youngest leaf of bananas by sending their germ tubes down through the stomata. Sporodochia develop in the cavities just under the stomata in about one month's time, and about 2 weeks later, conidia are produced on the sporodochia.

The best conditions for disease development are daily showers and night temperatures above 70°F. Below 70°, germ tube growth and infection are slow.

Day temperatures between 80° and 90° are more favourable for disease than temperatures between 70° and 80°. The disease is definitely worst on the wet sides of Viti Levu, Vanua Levu, Taveuni, and on the island of Vava'u in Tonga, than on the dry sides of the Fiji Islands and on Ha'apai and Tongatapu. In Samoa it is worse at sea level than at the higher altitudes.

The disease is also very bad on highly acid soils low in fertility.

Control

1. Avoid marginal soils for intensive banana culture. Maintain soil fertility in accordance with the results of soil tests and the recommendations of the Soil Chemist.
2. Through the wet, warm season from October to May, apply 2½ gallons of fungicide-oil-water emulsion per acre by misting machine every 10 days, or 3 times a month. Mix 2 lb maneb (Dithane M-22) or mancozeb (Dithane M-45), 1½ gallons misting oil, 1 gallon water, with 75 c.c. of Triton X-45 (Kemonic) *emulsifier,* added *first* to the oil.
3. During the dry, cool season from June to September, apply 2 gallons of fungicide-oil-water emulsion per acre every 14 days. Mix 1½ lb maneb (Dithane M-22) or mancozeb (Dithane M-45), 1 gallon misting oil, 1 gallon water, with 50 c.c. of Triton X-45 (Kemonic) emulsifier, added *first* to the oil. In localities without clearly defined warm and cool seasons, adjust spraying cycles to local temperature and rainfall conditions and to the results of local spraying trials. In rainfall belts of 70 inches per annum or less, the application of 1½–1 gallon of misting oil alone per acre at 2–3 weeks intervals may be sufficient.
4. For aerial application, mix 2 lb maneb (Dithane M-22) or mancozeb (Dithane M-45), 1–½ gallon misting oil, 5 gallons water, with 2 oz Triton X-114 emulsifier added *first* to the oil. Apply by aircraft every 10-14 days, depending upon local weather conditions.*
5. Measure the area to be misted accurately and make sure that the spray man or aircraft pilot travels at the right speed to deliver the right amount of emulsion or oil per acre. When spraying by misting machine, walk between the rows and do not aim the mist at the leaves or bunches. Use a machine capable of sending the mist directly upwards as high as the youngest leaves on tall plants. These leaves must be protected.
6. If maneb is not available, and minor leaf-spotting diseases, e.g., Malayan leaf spot, Cordana, or speckle, are very severe, add ½ pint of copper-in-oil (Cuproil) to each tankful of misting oil when spraying with oil alone.
7. Start misting new plantations about 2½ months after planting.
8. Keep misting machines in good condition and replace worn-out nozzles every 3 months.
9. Cut off dead leaves and *burn* them.

References: 9, 13, 31, 43, 48, 49, 76, 85.

*Recommendations for aerial application of maneb were supplied by the Rohm and Haas Company, Ltd., Philadelphia, Pa., U.S.A.

SPECKLE

Chloridium musae

Speckle is not important in Fiji and Tonga, but it is very obvious in both Western and American Samoa. In Fiji it is more noticeable in the Waidina Valley than elsewhere. Leaves are rarely killed by speckle but there may be some reduction in yield.

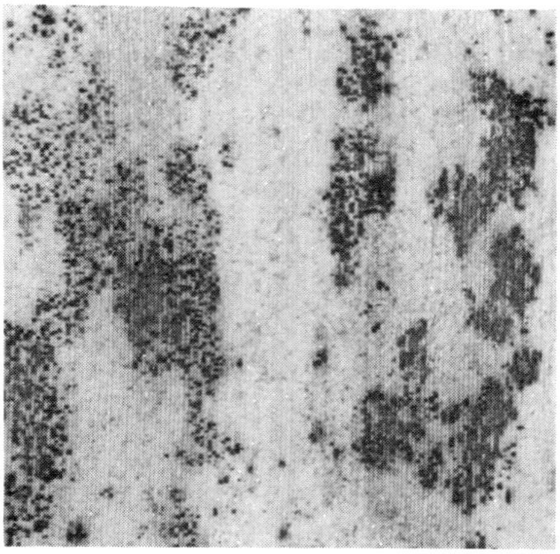

Fig. 43. Severe speckle on Veimama.

Symptoms

Leaves are covered with small, evenly spaced aggregations of black spots (Fig. 43) that are individually about the size of a pinhead. On the older leaves the dots may merge to form blotches.

Disease Cycle

Speckle may be caused by at least 3 different fungi, of which the most common is *Chloridium musae.* The disease needs further study in the South Pacific. Spores are produced on the undersides of banana leaves. They are spread by wind and rain to healthy leaves. Speckle appears to be favoured by shade, close planting, cool weather, and a very high rainfall.

Control

1. The winter schedule of maneb-oil-water emulsion recommended for control of Malayan Leaf Spot should also control speckle.
2. Avoid excessive shade (under coconuts) and too close a planting distance.
3. Prune the plants when they require it.

References: 56, 65, 85.

SPRAY INJURY

Improper Use of Chemicals

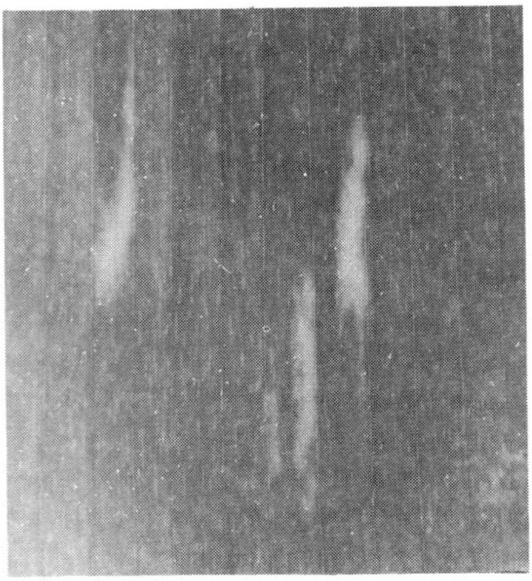

Fig. 44. Oil burn under dry weather conditions.

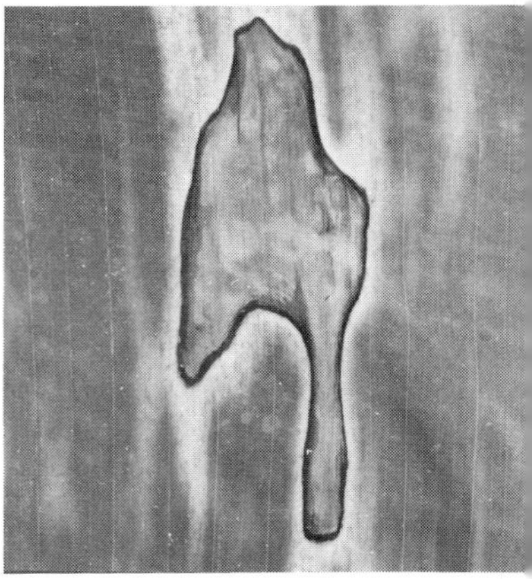

Fig. 45. Oil + maneb burn under dry weather conditions.

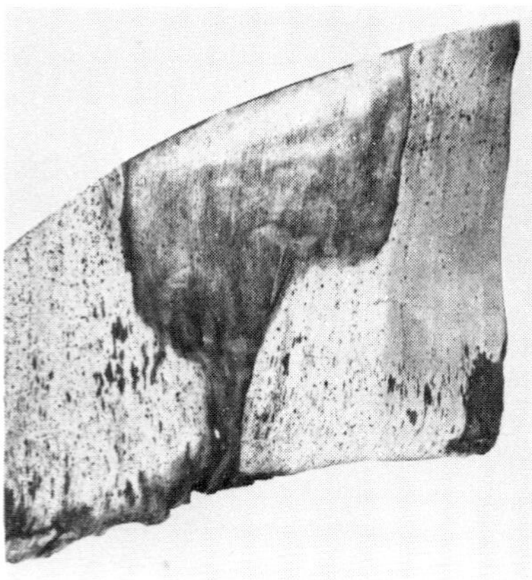

Fig. 46. Copper-in-oil burn caused by direct blast of spray.

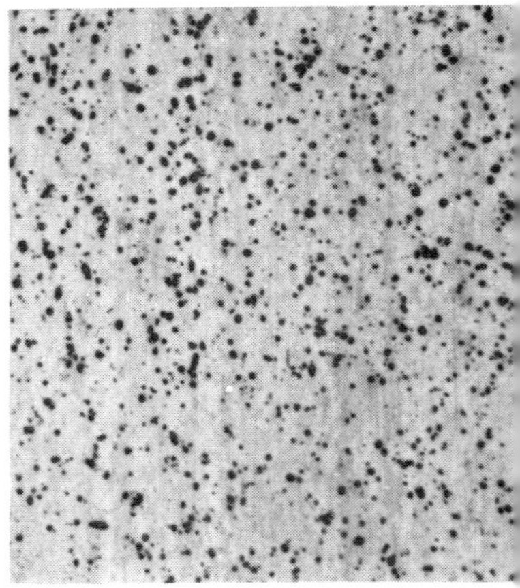

Fig. 47. Copper-in-oil injury under dry, cool weather conditions.

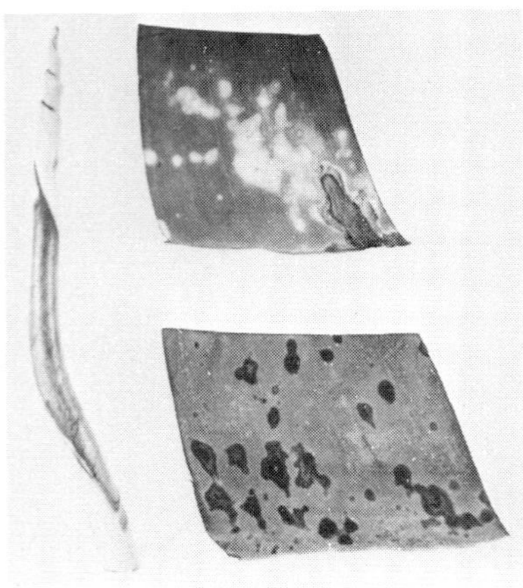

Fig. 48. Paraquat damage.

Fig. 49. Hand damage by direct blast of maneb-in-oil.

Intensification of banana culture in the South Pacific has increased interest in spraying for control of black leaf streak, Sigatoka disease, and other fungus diseases. Under certain conditions, the continuous use of misting oil alone and in combination with fungicides has caused injury to leaves and fruit.

Symptoms and Cause

(1) *Oil burn* – The continuous use of misting oil at the rate of 2 gallons per acre per fortnight causes a depression in growth rate and yield. Under very dry weather conditions, it burns the leaves. The injury is manifested as yellow streaks of varying size (Fig. 44).

(2) *Maneb + oil burn* – Maneb (Dithane M-22), now used in Fiji to control Malayan leaf spot, Cordana, and speckle during the cool season causes some burning under very dry conditions (Fig. 45). Fruit may be burned and blistered (Fig. 49), if the spray is blasted directly on to it.

(3) *Copper + oil burn* – Copper oxychloride and misting oil cause injury under humid conditions. Leaves are burned by a direct blast of the spray (Fig. 46). Under dry conditions, copper-in-oil injury resembles speckle (Fig. 47).

(4) *Herbicide damage* – The accidental use of knapsack sprayers and misting machines that have contained herbicide will cause severe injury. The hormone-type chemicals may cause twisting, scarring, and cracking of the fruit, and an outward curling of the edges of the unfurling leaves (Fig. 6). Paraquat causes injury resembling Cercospora leaf spot on young banana leaves (Fig. 48).

Control

1. Spray only with the recommended amounts of misting oil. During dry, cool weather reduce the amount applied from 2 gallons per acre to 1½ gallons per acre per fortnight. In low rainfall belts, frequency of spraying may be reduced to once every 3 weeks. Ensure that nozzle apertures are correct by replacing them when worn down.
2. Do not direct misting oil or combined sprays of oil and fungicide directly at the leaves or bunches. Walk directly between the rows and aim the mist upwards. Use an 11 x 5 spacing that will permit proper dispersal of spray.
3. Reduce the amount of maneb from 2 lb to 1½ lb per acre during dry cool weather, if the oil and fungicide combination is being used continuously.
4. If copper-in-oil is used for control of minor leaf spots, do not use more than ½ pint per 1½–2 gallons of misting oil. Avoid blasting the mixture directly on the leaves or fruit.
5. Use separate sprayers for application of herbicides.

References: 36, 43, 49, 85.

EYE SPOT

Curvularia sp.

Eye spot is common the year round on Batiki bluegrass but the grass is never killed. The disease appears to be worst at the beginning of the dry season when the grass enters a semi-dormant state. It is also severe on grass growing on hill soils lacking fertility.

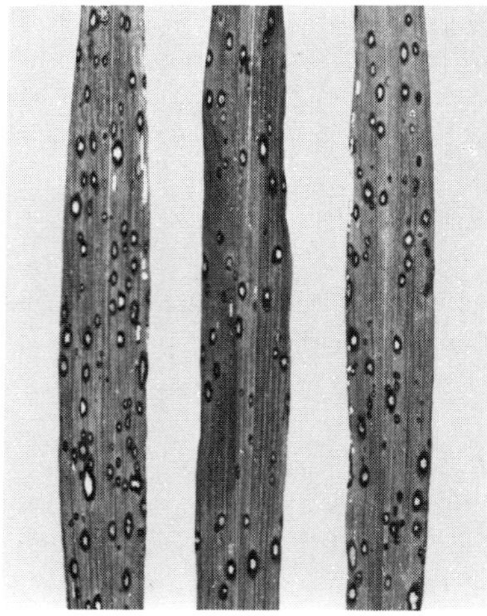

Fig. 50. Eye spot on Batiki bluegrass.

Symptoms

Small grey spots with purple borders cover the older leaves (Fig. 50). Some defoliation may occur.

Disease Cycle

The eye spot fungus is a species of *Curvularia,* a very common cause of leaf spot on many grasses. The disease appears worst when grasses are weakened by some environmental factor, or by old age. Dead spots produce *conidia* on long, brown *conidiophores* and these spores spread the disease.

Control

No special measures are recommended.

References: 24.

ROOT AND COLLAR ROT

Phytophthora nicotianae var. *parasitica*

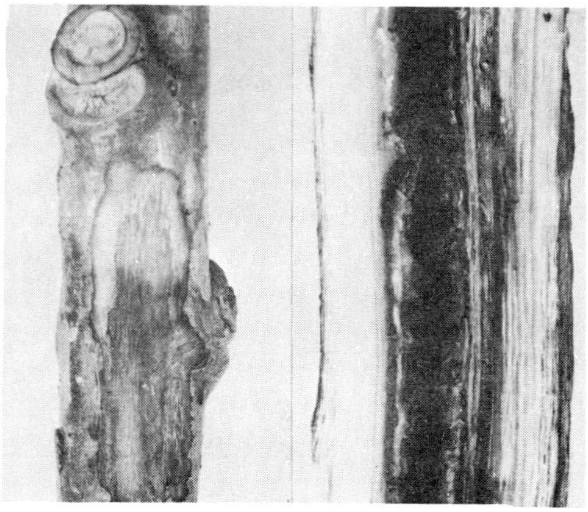

Fig. 51. *Left:* Bele stem cut to show advancing edge of invasion by the rot fungus; *right:* bele stem split to show rotted pith.

A severe outbreak of root and collar rot was found in a new plantation of bele at the Naboro Prison Farm in July, 1967.

Symptoms

Leave wilt and drop off the stems, commencing from the base and progressing upwards. Tissues at the soil line are severely decomposed. The advancing edge of the rot is watersoaked and reddish. If the outer tissues are cut away, a firm, reddish-brown rot is evident (Fig. 51).

If the stem is split the rotten wood and pith are exposed (Fig. 51). There is no recovery. Roots are also severely rotted.

Disease Cycle

As far as is known, bele *(Hibiscus manihot)* is a new host for *Phytophthora nicotianae* var. *parasitica.* Okva *(Hibiscus esculentus)* is a known host for this fungus in South-East Asia. This fungus is well known as the cause of fruit and root rot of papaw, dalo corm rot, passion fruit collar rot, pineapple heart and root rot, and probably damping-off.

The conditions favouring bele rot appear to be the same as for the other diseases.

Control

1. Select planting material with no symptoms of rotting.
2. Choose well-drained planting sites with good movement of air. Plant on mounds in very wet areas.
3. Avoid land previously planted to papaw, pineapple, passion fruit, okva and dalo, particularly if these crops have had the same trouble.
4. Remove and burn diseased stems as soon as they are detected.

References: 7, 35, 44, 65.

BLACK ROT

Xanthomonas campestris

Fig. 52. Wilted V's on the edge of a leaf.

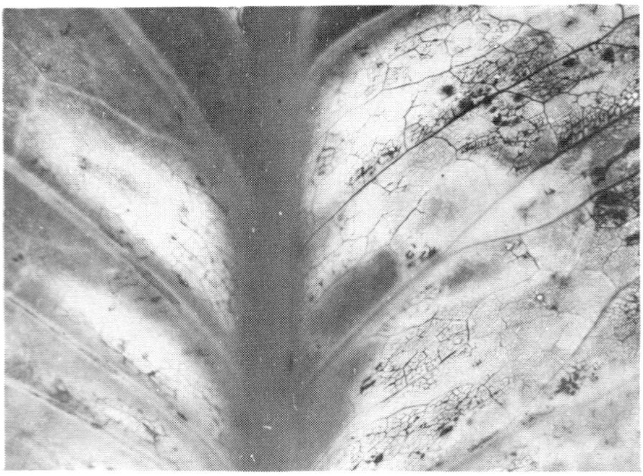

Fig. 53. Blackened veins surrounded by yellowed areas.

Black rot is unquestionably the worst disease of cabbage in the South Pacific. It has been found in Fiji, Tonga, New Caledonia, Norfolk Island, and Papua New-Guinea. Affected heads are often completely destroyed.

Symptoms

On seedlings the cotyledons turn black along the edges and fall off. On more mature plants small wilted areas in the shape of a "V" appear on the edges of the leaves (Fig. 52). The small veins at any point on the leaf surface may turn black (Fig. 53) and the affected area will then turn yellow. If the petioles or the stems are cut transversely, a black ring of vascular tissue may be seen. If soft rot follows black rot, the heads become a stinking mess.

Disease Cycle

The bacterium causing black rot is seedborne and it has probably entered Fiji and other islands repeatedly from such infested areas as New Zealand and Japan.

When the seed germinates, the bacteria infect the cotyledons and they invade the vascular system. Then they pass into the true leaves. They may be spread from plant to plant by insects, splashing and running water, and soil.

They can infect plants through stomata and hydathodes. The diease may be carried on weeds related to cabbage, such as mustard *(Brassica juncea, B. nigra)* and in soil between crops for at least one year.

Control

1. Obtain seed from seed-producing areas with low rainfall and insist on a certificate of freedom from black rot.
2. Treat doubtful seed with hot water at 122°F. for 15 minutes.
3. Use a 2-year rotation.

References: 14, 56, 70.

HOLLOW STEM

Boron deficiency

Hollow stem is common in cabbage growing on hill soils in the Suva Peninsula of Fiji. An associated soft rot causes considerable losses during wet weather.

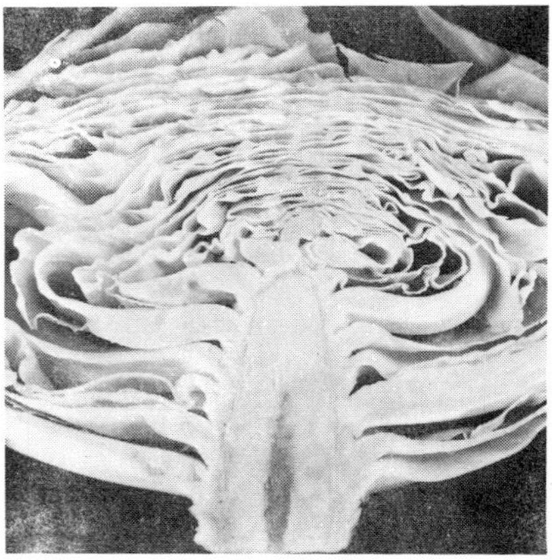

Fig. 54. Softened pith in boron-deficient cabbage.

Symptoms
Cabbage stems are hollow or with a soft, discoloured pith (Fig. 54). Soft-rotting may follow to turn heads into a stinking mess. Heads are undersized and the petioles are brittle.

Cause
Boron is deficient in alkaline soapstone soils. This element is apparently essential for cell-wall formation. Without it, cell walls are thin and tissues collapse. Conducting tissues may also be affected. All vegetables related to cabbage, e.g., cauliflower, brussels sprouts, broccoli, Chinese cabbage, suffer from boron deficiency. The weakened tissues may be invaded by soft-rot bacteria.

Control
1. Apply powdered borax with the fertilizer at the rate of 30 lb per acre.

References: 14, 84.

ALTERNARIA BLIGHT

Alternaria dauci

Alternaria leaf blight has been reported in Fiji, New Caledonia, and Papua-New Guinea. It can cause severe loss of leaves and a reduced crop.

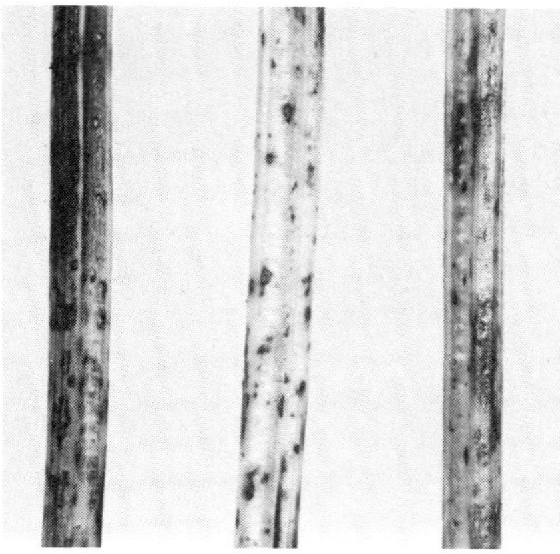

Fig. 55. Alternaria blight of carrot petioles.

Symptoms

Small dark-brown to black spots with yellow margins form on the leaves. The spots grow larger and eventually run together to kill the leaves, which then turn black. Elongated rust-coloured spots (Fig. 55) appear on the petioles. Under certain conditions, damping-off of carrot seedlings and a black root rot may occur.

Disease Cycle

The fungus is carried on the seed and also in diseased refuse in the soil. The brown spores *(conidia)* that are produced on the blighted leaves and petioles are spread by running water, splashing rain, contaminated soil, and machinery. Infection is favoured by wounds but otherwise it is very slow. New spores are produced about 2 to 3 weeks after the first symptoms appear.

Best conditions for disease development are air temperatures around 80°F. and frequent rains.

Control

1. Plant carrots on well-drained soil and use a 3-year rotation.
2. Dust seed with captan (1½ tsp/1 lb).

References: 14.

ROOT KNOT

Meloidogyne incognita

Carrots grown in the Suva Peninsula of Fiji show a high rate of infestation of root knot nematodes. The species is probably widespread in the South Pacific. Root knot disfigures the roots but it may not reduce yield by very much. Imported carrots are purchased in preference to local ones.

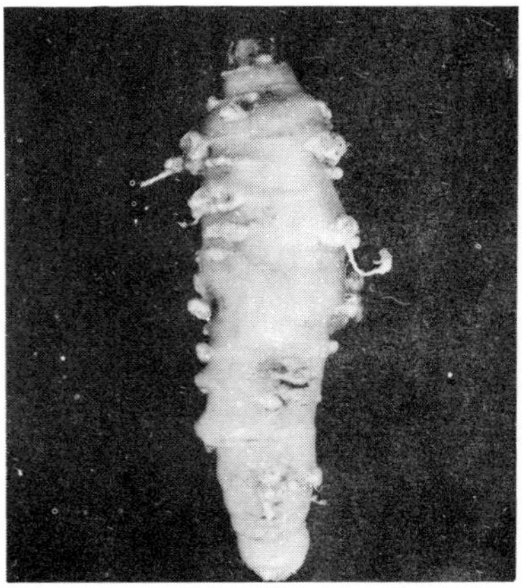

Fig. 56. Root knot infestation of carrot.

Symptoms

The main roots and laterals are covered by swellings that vary in size from ⅛-½ inch in diameter (Fig. 56). Infested carrots may be reduced in size. The plants may be stunted, yellowed, and slightly wilted during dry weather.

Disease Cycle

Root knot is caused by the same nematode that attacks bananas, tomatoes, tobacco, ginger, and many other crops and weeds.

Eggs hatch to produce larvae that enter root tips. Inside the root cortex they moult a number of times before becoming eelshaped males or pearshaped females. The mature gravid (pregnant) female can be seen as a glistening white pinhead if the galls are cut open. As the female grows, the surrounding tissues of the root are stimulated to produce galls. Eggs are laid in a jelly-like mass that breaks through the root surface. The eggs are then released into the soil. The length of the life cycle varies between 3 weeks and 3 months, depending upon the host plant and environmental conditions, such as soil moisture and temperature.

Root knot nematodes are spread by infested planting material, implements, vegetable refuse, and to some extent, by flowing water. On their own, they can move only a few feet a year.

Control

1. Rotate carrots with rootknot-nematode-tolerant crops such as rice, maize, cabbage, and avoid susceptible pulses and solanaceous crops.
2. Keep weeds under control, particularly those belonging to the Solanaceae.
3. Plant carrots in the cool season when nematode activity is reduced.
4. Fumigate the soil at least 2 weeks before planting with EDB (ethylene dibromide) emulsion at the rate of 10 gallons per acre on a 12-inch spacing. Cultivate the soil to a very fine state before fumigation.

References: 18, 32, 36, 78.

LEAF SPOT

Cercospora bradburyae

Leaf spot has been found on centro *(Centrosema pubescens)* wherever this forage legume grows in Fiji. It is not important.

Symptoms

Black, starlike spots surrounded by yellowish haloes, about ⅛-¼ inch in diameter, appear mainly on the older leaves (Fig. 57). If the disease is severe, leaves turn yellow and drop.

Fig. 57. Leaf spot of centro.

Disease Cycle

The disease is spread by *conidia* that are produced in clusters on *stromata* found usually on the upper surfaces of spots. It is quite possible that *Cercospora bradburyae* can infect other legumes. Details of the life cycle are similar to those of other fungi in the same group.

Best conditions for leaf spot are found in shaded locations where dew is persistent.

Control

No special measures are recommended.

References: 13.

MOSAIC

Cucumber Mosaic Virus

Mosaic is quite common in chillis, wherever they are grown. It is probably responsible for a great deal of reduction in plant size and yield.

Fig. 58. Vein-clearing on terminal leaves of chilli.

Symptoms
Usually the first symptom to show is vein-clearing or the development of yellow bands along the veins (Fig. 58). Leaves become mottled light and dark green. They are smaller and narrower than normal healthy leaves. If the plants are infected when they are quite young, they are stunted and bushy. Fruit of the large-fruited varieties is reduced in size and warty in appearance.

Disease Cycle
The cucumber mosaic virus can be transmitted by aphids or by contact. Workers handling seedlings transmit the virus from plant to plant by the sap that gets on their hands. It may be present in weeds in and around chilli fields.

Control
1. Keep weeds, particularly the solanaceous ones, under control.
2. Be very careful in handling seedlings. Wash the hands with strong soap and water containing 0.5% Teepol or Byprox.
3. Rogue out infected plants as soon as they are detected.

References: 11, 14, 74.

DOWNY MILDEW

Peronospora parasitica

Fig. 59. Symptoms of downy mildew on upper leaf surface of Chinese cabbage.

Fig. 60. Symptoms of downy mildew on under side of Chinese cabbage leaf.

Downy mildew was observed on Chinese cabbage for the first time in Fiji at the Naboro Prison Farm. A mild infestation was found in about 100 plants. It is impossible to predict its effect on the crop as a whole.

Symptoms

An interveinal yellowing similar to that caused by certain nutrient deficiencies appears on the upper sides of leaves. The yellowed areas turn to brown (Fig. 59) and the leaf becomes papery in texture. Lower leaves may fall off. Lower sides of leaves are covered by a white powdery mould (Fig. 60).

Disease Cycle

The downy mildew fungus is worldwide in distribution but the strain on Chinese cabbage may be different from that on other Brassica species, e.g., cabbage, cauliflower, radish, turnip. These crops are not affected by downy mildew in Fiji. The fungus is spread by seed from one country to another. The *conidia* (or *conidiosporangia*), which are produced in great numbers on the undersides of leaves, spread the diease within the crop. These spores germinate directly by germ tubes, which infect the leaves and produce a new crop of spores in less than a week. The fungus survives between crops on old diseased plant parts, as thick-walled *oospores*.

Best conditions for downy mildew are cloudy weather and a sheltered planting location where dew persists for most of the day. The most favourable night temperatures are between 45° and 60°F., with day temperatures round 75°F.

In the United States the P/K ratio has been found to influence the severity of downy mildew to a significant degree. An imbalance in favour of potash is supposed to favour the disease.

Control

1. Pull and burn diseased plants as soon as they are detected.
2. Treat seed with hot water, as for control of black rot.

3. Spray the seedbed with spergon (Coversan) (2 oz/3 gallons) 2 or 3 times a week. Continue the treatment until the seedlings are ready to transplant. Spray plants weekly in the field only if the disease is consistently severe. Do not spray within 2 weeks of harvest.
4. Check the soil fertility.

References: 14, 62.

LEAF ROT

Corticium sasakii

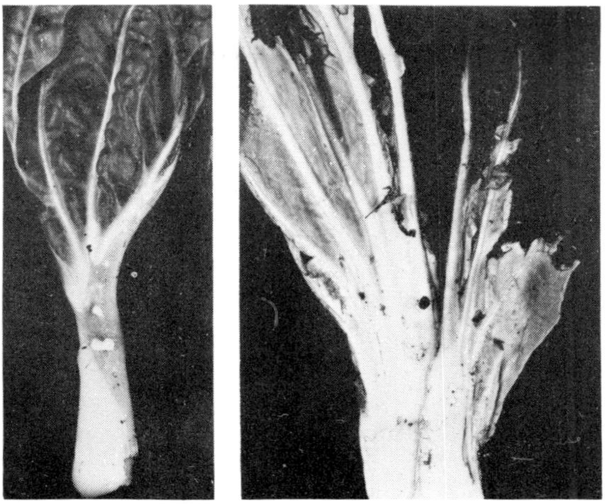

Fig. 61. *Left:* White sclerotium on a petiole; *right*: a shredded leaf.

Leaf rot destroys a small number of plants every year but it is serious only after floods or after repeated planting on rice land.

Symptoms

Leaves are shredded as if they had been attacked by chewing insects. Sometimes white or greenish-brown cushions about half the size of a pea can be seen on the petioles and leaves (Fig. 61). If the affected areas are examined closely, a white cobwebby fungus growth can be seen. Soft rot often follows an attack of leaf rot.

Disease Cycle

Leaf rot is caused by the rice sheath rot fungus, *Corticium sasakii.* It produces *sclerotia* or cushion-like lumps of resistant tissue that can live for very long periods in the soil. The fungus grows out from these lumps or from *mycelium* in undecayed crop refuse on to healthy plants. It is spread by running water and by spores produced on the surfaces of leaves and stems. Best conditions for disease development appear to be close planting and very moist soil.

The leaf rot fungus attacks many hosts, including cabbage, cowpeas, rice, grasses, and water hyacinth.

Control

1. Plough in crop residues deeply (9-10 inches)
2. Do not plant Chinese cabbage too closely. The leaves of neighbouring plants should not touch each other.
3. Avoid too much nitrogen in the fertilizer.
4. Spray young plants with Kopi or Cuprox at the rate of 2 oz/3 gallons.

References: 60.

MOSAIC and YELLOWS

Virus

Fig. 62. Symptoms of mosaic.

Fig. 63. Symptoms of yellows.

Mosaic and yellows often occur in Chinese cabbage when aphid infestations are high, and can seriously reduce the yield.

Symptoms

A mosaic pattern of light and dark green areas appears on the leaves. The dark green parts may be raised like blisters (Fig. 62). Sometimes there are broad bands of yellow or white on the outer leaves (Fig. 63). The plants are reduced in size.

Disease Cycle

"Although no experimental work has been done locally on this disease, it is suspected to be a strain of pechay mosaic, found on Chinese cabbage in the Philippines. There it is transmitted by the maize aphid *(Aphis maidis)* and the melon aphid *(Aphis gossypii)*, both of which occur in Fiji on Chinese cabbage. Weeds such as mustard *(Brassica nigra)* are suspected to be alternate hosts for the virus."

The symptoms do not fit those described for Chinese cabbage or cauliflower mosaic *(Brassica Virus* 3*)*, a common disease on crucifers in temperate-zone countries. The host range also appears to be different because mosaic has not been observed locally on cabbage.

Control

1. Rogue affected plants as soon as they are detected.
2. Spray the crop, particularly during the seedling stage, with malathion (1 oz/3 gallons).

References: 14, 23A, 74.

SOFT ROT

Erwinia spp.

Soft rot is common on Chinese cabbage in Fiji, but it rarely causes serious losses.

Fig. 64. Soft rot and hollow stem of Chinese cabbage.

Symptoms
The central leaves of the plants wilt. They are rotted usually from the base upwards (Fig. 64). Stems are hollow. The rotted parts may be invaded by maggots.

Disease Cycle
Soft-rotting bacteria are the cause but they are not primary. It is possible that the hollow-stem condition found associated with soft rot is caused by a deficiency of boron or of some other minor element.

Control
1. Check soil fertility and acidity.
2. Apply borax with the fertilizer at the rate of 30 lb per acre.

References: 14, 84.

WHITE RUST

Albugo candida

White rust is common in Fiji on Chinese cabbage during the cool season. It does not attack cabbage. It has very little effect on yield but it makes the leaves unsightly.

Fig. 65. White rust pustules on Chinese cabbage.

Symptoms
Pustules form on the leaves and petioles. They break open to expose a white powdery mass (Fig. 65) in circular or elongated cavities. Leaves may be distorted.

Disease Cycle
The white rust fungus is common on cruciferous weeds closely related to Chinese cabbage. The most common of these are the mustards, *Brassica nigra* and *B. juncea.* The white powdery mass inside the pustules consits of sporangia, that are blown by wind to neighbouring healthy plants. They germinate in water to produce swimming spores *(zoospores),* that swim about in droplets on the leaves. Then they settle down, germinate, and penetrate the leaves when they are very young. Only the tissues just underneath the epidermis are infected. Under certain conditions, usually very cold weather, resistant spores *(oospores)* form in affected tissues. It is unlikely that these spores are formed under tropical conditions, because the weather the year round is favourable for production of sporangia and zoospores.

Best conditions for disease development are temperatures around 68°F., intermittent rain, and an abundance of susceptible hosts. The green varieties of Chinese cabbage are more susceptible to rust than the purple ones.

Control
1. Pull and burn severely diseased plants.
2. Rotate cabbage with non-cruciferous crops and keep weeds under control.
3. Spray young plants with spergon (Coversan) (2 oz/3 gallons).

References: 14, 62.

ANTHRACNOSE

Glomerella cingulata, Gloeosporium limetticola

Anthracnose is most serious on Mexican lime, on which it is also known as "wither tip". Under certain conditions, it affects the ripening fruit of Lisbon lemon and grapefruit, and the foliage of bush lemon.

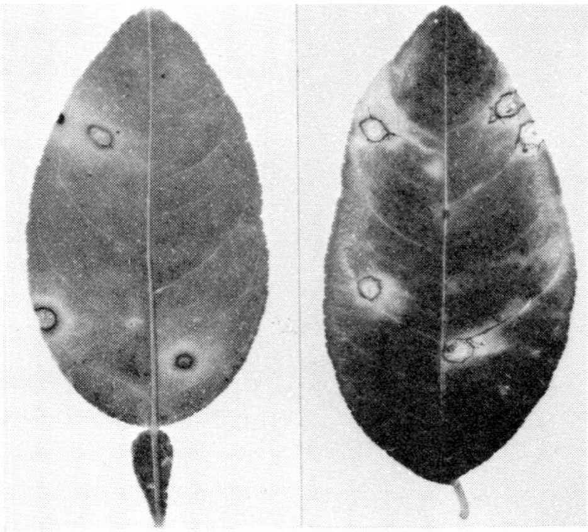

Fig. 66. *Left:* Anthracnose on Mexican lime; *right:* anthracnose on bush lemon.

Symptoms

Twigs die back from the tips and turn white. Gumming may occur at the juncture between diseased and healthy tissue. The whitened bark is studded with black pinpoints. Mexican lime leaves have round white spots about $\frac{1}{8}$-$\frac{1}{4}$ inch in diameter (Fig. 66), with black pinpoints in their centres. On bush lemon, the leaf spots are roughly eyeshaped, grey, and papery. Sometimes the centres of the spots fall out. In severe attacks, the leaves may fall off. Fruit of lemon and grapefruit develop black sunken spots that exude gum. There is no cork formation as with canker. Mexican lime fruit may split.

Disease Cycle

The disease on lime is generally considered to be caused by a different fungus species, *Gloeosporium limetticola.* For all practical purposes, anthracnose and wither tip are one and the same disease, and their cycles are essentially the same. Spores are produced in great masses on the black pinpoints *(acervuli)* on twigs, leaves, and fruit during rainy weather and they are blown to healthy trees. The sites of infection are mechanical wounds or weakened tissues.

Any condition causing wounds or weakened tissue, such as poor nutrition, wind damage, pruning cuts, insect attacks, encourages anthracnose.

Control

1. Check soil fertility.
2. Remove dead wood and treat pruning cuts with a tree protectant.
3. Pick the fruit before it is overmature.
4. Spray nurseries with captan (1 oz/3 gallons).

References: 29, 45, 47.

Unknown

Fig. 67. Dieback of Lisbon lemon after severe bark cracking.

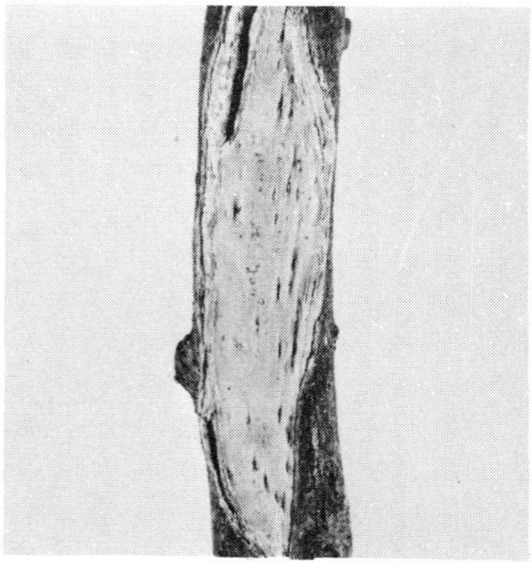

Fig 68. Bark cracking and stem pitting.

A survey of Fiji citrus diseases in 1965 has shown that almost all Lisbon lemons, seedless limes, grapefruit, and some sweet oranges grafted on to bush lemon stocks suffer from bark cracking. Mandarins and sour oranges are relatively free. Little or no cracking was observed on seedling trees of sweet orange and mandarin, but it was found on some seedling bush lemons.

Symptoms

The bark cracks lengthwise from the bud union up into the crown of the tree. Cracking may be accompanied by production of gum. Trees over 7 years old tend to lose their leaves and die back (Fig. 67). Root systems appear healthy. Sometimes the exposed wood of Lisbon lemon and bush lemon shows severe pitting (Fig. 68).

Cause

The exact cause is unknown. White scale is often found associated with bark cracking, but not consistently so. The opinion has been offered by visiting citrus specialists that it is caused by the stem-pitting virus (xyloporosis). This can be proven only by grafting on indicator stocks such as citron and trifoliate orange.

Control

1. Keep scale insects under control according to the recommendations of the Entomologist.
2. Try rootstocks other than bush lemon, e.g., sweet orange.

References: 46, 63, 82.

BLUE and GREEN MOULD

Penicillium spp.

These common fruit rots occur everywhere on citrus after harvest. They are not important in the South Pacific because fruit is rarely held for very long or shipped over long distances.

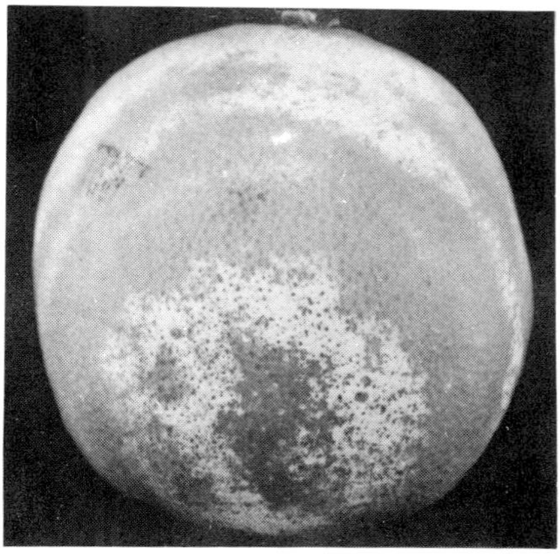

Fig. 69. Ripe fruit rot of orange.

Symptoms
Fruit is covered by a bluish or greenish mould with a white, powdery or pasty margin and a halo of water-soaked tissue (Fig. 69). The fruit eventually collapses into a watery mess.

Disease Cycle
The fungi that cause these rots are soilborne and airborne. They can invade only wounds, such as those caused by rough handling. They can also be spread from rotten to healthy fruit by contact.

Control
1. Pick fruit when it is dry and avoid rough handling.
2. Do not put fruit that has been on the ground in the same container as fruit picked off the tree.

References: 29, 45, 65.

Xanthomonas citri

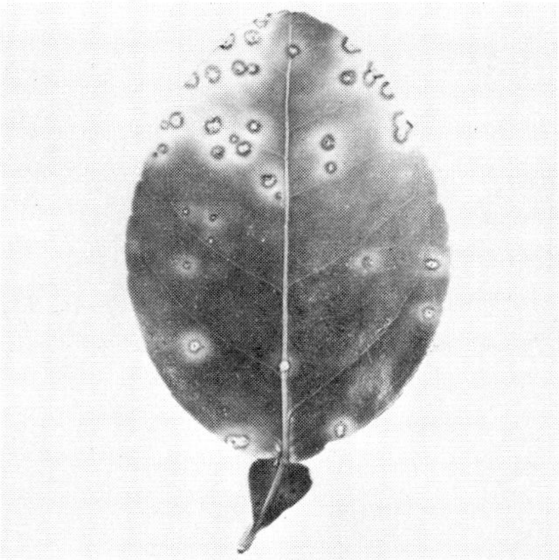

Fig. 70. Canker on leaf of grapefruit.

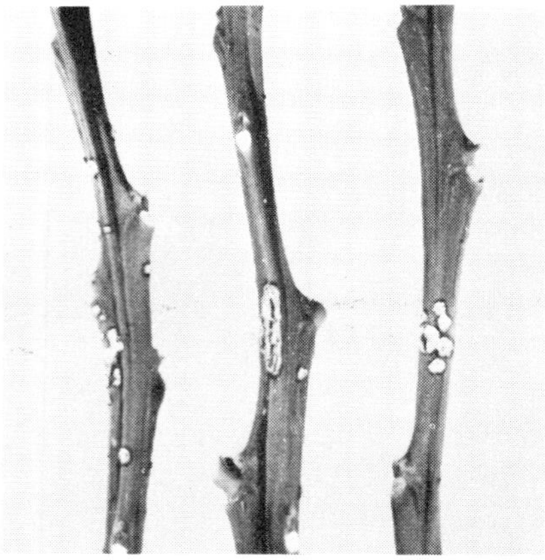

Fig. 71. Canker on twigs of grapefruit.

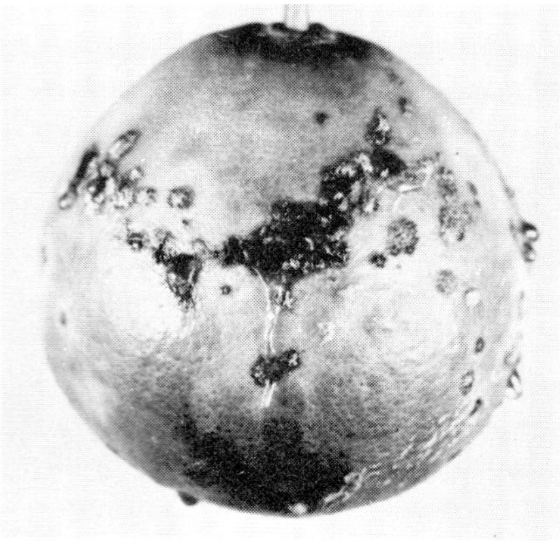

Fig. 72. Canker and exuding gum on fruit.

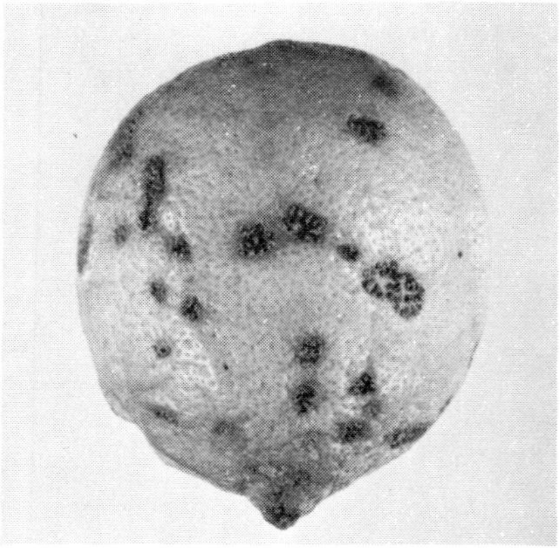

Fig. 73. Canker on Lisbon lemon.

Canker is unquestionably the worst disease of citrus. It has been reported in the South Pacific from Fiji and New Guinea, and from Guam and the Caroline Islands in the North Pacific. In Fiji it has prevented the development of a citrus industry. An attempt to eradicate the disease in 1953-55 seems to have been partially successful, because canker has not been observed in the Waidina and Sigatoka valleys, Vanua Levu, Taveuni, Rotuma, and many islands of the Lau group.

Grapefruit is very susceptible; sweet orange and Mexican lime are moderately susceptible; seedless lime, lemon, mandarin, and kumquat are commercially resistant.

Symptoms

Small, watery, translucent spots usually darker green than the surrounding tissues, and with convex surfaces, appear on the leaves. They occur more frequently on the undersides. The spots turn grey and rupture to form small craters exposing a light brown spongy depressed central mass. The spots are surrounded by yellow haloes (Fig. 70). The lesions on twigs (Fig. 71) and fruit (Fig. 73) are similar to those on the leaves but there is more cork formation.

On grapefruit and orange, the infected fruit may split and exude gum (Fig. 72). Lesions are largest, up to ½ inch in diameter, on grapefruit; intermediate, up to ¼ inch, on sweet orange; smaller on lemon and lime. The lesions often develop concentric rings of corky tissue. Grapefruit is often affected by leaf dropping and dieback.

Disease Cycle

Canker bacteria are exuded from spots by the millions every time it rains. The disease spreads slowly but it can flare up badly when rain is accompanied by winds of hurricane force. The water droplets carrying bacteria are splashed on to young leaves, fruit, and twigs. Invasion occurs through stomata, lenticels, or small wounds.

Animals, insects, budwood, and pruning tools can also transport the bacteria. After infection, the bacteria may remain inactive for a long time in lesions on the older bark. They die rapidly in soil, and as far as we know, they do not attack roots.

Control

1. On small islands, eradicate canker by cutting down and burning every citrus tree. Place a strict quarantine on trees and fruit coming from neighbouring infested areas. Do not replant for at least a year and use trees from uninfested areas.
2. If it is impractical to eradicate canker because of inaccessibility of terrain, do not plant highly susceptible grapefruit and oranges. Concentrate on lemons, mandarins, and seedless limes.

References: 26, 29, 57, 68.

FELT FUNGUS

Septobasidium separans

Fig. 74. *Left:* Felt fungus on grapefruit twig; *right:* felt fungus and scale insects on lemon fruit.

This condition does very little harm to citrus unless it grows over leaves and cuts out the light. It can also disfigure fruit.

Symptoms

A compact, light brown to grey felt-like mantle (Fig. 74) grows over twigs, petioles, and sometimes over the surface of leaves and fruit. On twigs it may extend for 1-3 inches.

Disease Cycle

The fungus is actually parasitic on colonies of scale insects on the surface of the bark, leaves, etc. It may be a form of biological control of scale. It develops best in very humid locations. Spread is accomplished by *basidiospores* that are produced on the felty surface and blown about.

Control

No special measures are needed other than to control scale.

References: 2, 29, 47.

FOOT ROT, GUMMOSIS, and BROWN ROT

Phytophthora nicotianae var. *parasitica*

Fig. 75. Foot rot of seedling bush lemons.

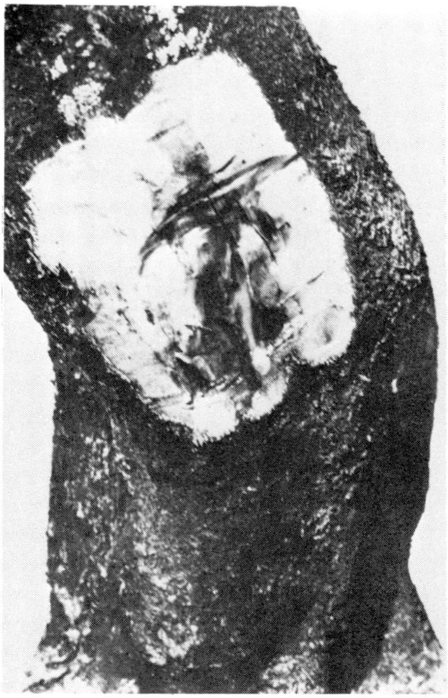

Fig. 76. Foot rot of mature citrus.

This disease complex is apparently not a serious problem in Fiji but it has been observed in the wet zones. The order of resistance to foot rot and gummosis appears to be: trifoliate orange, sour orange, mandarin, sweet orange, grapefruit, bush lemon, lime, and Lisbon lemon. Perhaps the local strains of sweet orange *(moli taiti)* are resistant, because seeding trees abound.

Symptoms

A watersoaked dark spot appears on the trunk, usually at the bud union or at soil level on seedling trees (Fig. 75). As the wound expands, the dead bark dries out and sloughs away in lengthwise strips. A clear, yellowish-brown gum oozes from the wound. Eventually, the tree is ringbarked and dies back, often in a one-sided manner, depending upon the extent of the wound. Roots are also invaded and develop a sour-smelling rot. If the rotted area is cut open, the underlying tissues show a reddish-brown discolouration extending across the cambium into the wood (Fig. 76).

Fruit may also become infected and it develops a firm, leathery, brown rot with a strong fermentative odour. On stored fruit a white fuzzy, mouldy growth may develop.

Disease Cycle

At least 3 species of *Phytophthora,* a water mould, are involved elsewhere in foot rot and brown rot. The one that has been positively identified in Fiji also attacks bele, dalo, papaw, pineapple, and possibly passion fruit, causing similar rots. Spores produced in the soil infect wounds in the bark when carried on to it by splashing rain. Uninjured bark may also become infected if it remains wet for long periods. Best conditions for infection are: (1) abundant soil moisture at ground level where bark is in contact with the soil; (2) soil and air temperatures of 60-70°F. at night, stimulating zoospore formation, and 80-90°F. in the daytime, favouring fungal growth; (3) bark wounds; (4) susceptible stocks.

Fruit is infected under the same favourable conditions by zoospores that are splashed on to the lower branches during periods when the fruit does not dry readily. Fallen fruit is infected by soilborne spores.

Control

1. Experiment with sweet orange as a rootstock in heavy wet soils.
2. Keep the bud union high (12-18 inches) to avoid splashing susceptible scions with water containing spores.
3. If at all possible, plant citrus in well-drained locations.
4. Keep weeds and grass cut away from the bases of trees.
5. If the tree is not completely ringbarked, pare away diseased bark, disinfect the wound with Clorox or Javex (1 part/10 parts of water), and paint it *after* it shows signs of healing with tar or a wood preservative.

References: 29, 45, 47, 89.

GREASY MELANOSE

Mycosphaerella horii

Greasy melanose is seen occasionally on lemons and grapefruit but it is only superficial.

Fig. 77. Greasy melanose on grapefruit leaf.

Symptoms

Yellowish-brown spots appear on the upper sides of leaves. On the lower sides, there are oily brown patches resembling grease (Fig. 77). Sometimes the fruit shows symptoms of grease spot. Under very favourable conditions, some defoliation may occur.

Disease Cycle

The life cycle of the fungus has not been completely worked out. There is evidence that infection is primarily from *ascospores* produced in *perithecia* in dead leaves. These are probably ejected during rain on to the lower surfaces of leaves.

Control

1. Spray with Cuprozin as for the control of scab and melanose.
2. Do not spray with Bordeaux mixture or Perenox. These copper compounds aggravate existing lesions of greasy melanose into a condition known as "star melanose".

References: 47.

HEART ROT

Ganoderma sp.

Fig. 78. Grapefruit trunk with heart rot at point of topworking.

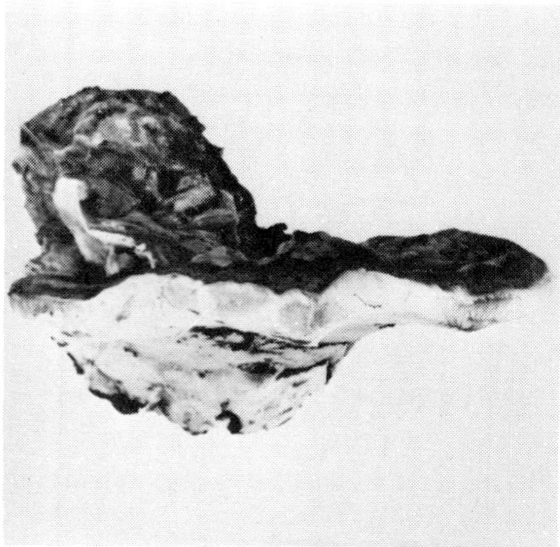

Fig. 79. Fruiting body of *Ganoderma* sp. (from coconut).

Heart rot is seen occasionally in citrus injured by pruning or topworking operations. It can also result from hurricane damage.

Symptoms

The first symptom is the presence of wood-eating ants *(Pheidole megacephala)* and sawdust around the trunks of affected trees. Branches that are badly rotted inside snap off easily in heavy winds. Dieback may also be evident.

Heart rot in the first stage causes a discoloration of the wood, which remains firm. In the final stages, it is soft and punky, and it may be invaded by boring insects.

The most obvious symptom is the presence of the fruiting bodies of bracket fungi (Fig. 78) which are capable of invading healthy wood. These may form at the point of injury or at any point along the invasion pathway.

Trees eventually die unless remedial measures are taken.

Disease Cycle

The starting points for heart rot are pruning cuts, splits in the trunk or branches, root injuries, sunscald, rat injury to bark, or topworking operations (Fig. 79). In fact, any injury to bark may serve as an entry point for heart-rotting fungi. One of the most common of these is *Ganoderma.* In the tropics, *Fomes noxius,* another bracket fungus, can also be associated with trunk and root rot. Heart rot may follow the attacks of other pathogens, e.g., *Phytophthora* species that cause foot rot and gummosis.

Control

1. Make all pruning cuts flush with the branch or trunk surfaces. Notch branches from below to prevent tearing and splitting of large cuts.
2. If dead wood is being pruned out, make cuts through healthy wood.
3. Treat cut surfaces with a wound protectant such as coal tar. Disinfect first by painting with a fungicide, e.g., Bordeaux paste.

References: 29, 47, 65.

LICHENS

Algae and Fungi

These growths are frequently mistaken for serious disease. They are an indication that citrus plantations are being neglected.

Symptoms
Scaly crusts grow on the leaves, branches, and twigs. They may be so numerous that they completely cover leaf surfaces (Fig 80). The colour of the crusts is quite variable.

If the growths are numerous, the trees may suffer some harm through exclusion of sunlight, or a plugging of the lenticels or breathing pores on the bark.

Fig. 80. Lichens covering grapefruit leaf.

Cause
The growths are an association between certain algae and fungi together. The algae use the fungi for support, while the fungi break down some of the organic matter in bark and make the nutrients available to the algae. The algae photosynthesize and supply the fungi with sugars. The growths are not parasitic on leaves or bark.

Control
1. Keep trees well pruned and fertilized.
2. Spray with Cuprozin as for control of scab and melanose.
3. If the infestation is very severe, spray with Bordeaux mixture (3-4-50).

References: 47, 65.

LITTLE LEAF

Zinc deficiency

This disorder is fairly common on extremely alkaline and acid soils. It affects all varieties of citrus. Oranges are most severely affected.

Fig. 81. Little leaf of orange.

Symptoms

The leaf tissue between the veins turns yellowish-white, while that adjoining the veins remains green. New leaves are undersized, narrow, and more acutely pointed (Fig. 81). Twigs are small, short and bushy, and they tend to die back. Water sprouts may develop in the interior of the tree. Fruit is small, with woody pulp and a sour taste. If fruit is produced on water sprouts, it is large and coarse.

Cause

Little leaf is generally attributed to a deficiency of zinc, which becomes unavailable in extremely alkaline and acid soils.

Control

1. As a corrective spray, apply zinc sulphate to the foliage at the rate of 1 oz/3 gallons. For maintenance, apply a spray of ½ oz/3 gallons.
2. Where Cuprozin or zineb is being used continuously for control of fungus diseases, zinc deficiency is probably corrected automatically.
3. If practical, bring the soil pH close to 6.0 by the addition of acid fertilizer or lime.

Refrence: 8, 65.

MELANOSE

Diaporthe citri

Melanose is common on all varieties of citrus in high rainfall areas in Fiji. It has also been reported in Papua-New Guinea. It has little effect on fruit quality other than to spoil its appearance.

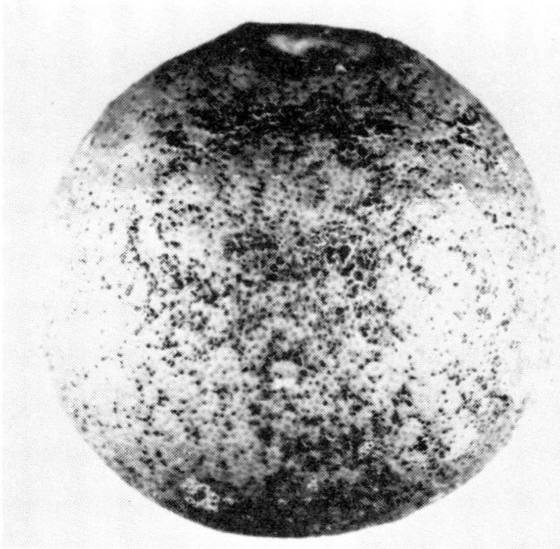

Fig. 82. Melanose on sweet orange.

Symptoms

Raised dots of pinpoint size (Fig. 82) appear on twigs, leaves, and fruit, making their surfaces feel like sandpaper. If infection is very severe, continous crusts or "mud-cakes" develop on fruit. Sometimes a tear-stain pattern appears if spores are carried down the sides of fruit in water droplets. A severe attack on the pedicels may cause stem-end rot after harvest.

Disease Cycle

Conidia are produced in great numbers in the *pycnidia* on dead twigs and leaves. When it rains the spores are pushed out of the pycnidia in tendrils and they are washed by rain water over new leaves and young fruit. These spores infect within 36 hours after they are deposited, and visible spots appear about one week later. *Perithecia* may also be produced on dead twigs and branches, but these are relatively unimportant in the spread of the disease.

Best conditions for disease development are a high humidity and temperatures between 70 and 80°F. Any factor that increases the amount of dead wood, e.g., drought, bad drainage, hurricane, insect attack, will increase the amount of melanose.

Control

1. Prune and burn all dead wood.
2. Spray with Cuprozin, as for control of scab.

References: 45, 47, 65.

Tristeza Virus

Fig. 83. Mandarin showing symptoms of quick decline.

Fig. 84. Vein clearing indicating presence of tristeza virus in Mexican lime.

Quick decline was identified in Fiji in 1965 on seedless limes, Mexican limes, and kumquats. It is probably quite common everywhere on citrus but it may be relatively unimportant because the popular bush lemon rootstock is tolerant.

Symptoms

Quick decline resembles any condition caused by root injury, ringbarking, bad drainage, foot rot, or excessive fertilizer. Leaves turn grey, roll upwards, and dieback (Fig. 83) develops. Meyer lemons usually come into bearing prematurely. Mexican limes, seedless limes, and kumquats show the characteristic vein-clearing symptom (Fig. 84). On sour orange stocks infected buds fail to grow or the scions produce stunted trees.

Stem pitting develops on grapefruit, limes, and sour oranges. If the bark is peeled back, there is a honeycombing consisting of closely spaced fine holes on the inner bark surface. This symptom is most evident on sour orange rootstocks just above the bud union.

Small fibrous roots die and the bark sloughs off. Eventually larger roots die and dieback develops. Trees may tolerate tristeza virus infection for 10-15 years while remaining relatively unproductive. Once decline starts the trees may be dead within 3-6 months.

Disease Cycle

Tristeza virus is transmitted in Fiji by the black citrus aphid, *Toxoptera citricidus,* which is most abundant on citrus. The virus can also be carried and spread in infected budwood. The aphids pick up the virus while feeding on the very young tissues of diseased trees. Winged forms develop that fly away to feed on other trees. The virus is injected into healthy trees when the insect is feeding. The aphids are phloem feeders and therefore the virus moves into and through this tissue to all parts of the tree, including the roots. Here it destroys the food-conducting vessels and produce the stem-pitting and furrowed appearance of the inner bark and the ringbarking effect.

Control

1. Use of the tristeza-tolerant rootstocks bush lemon and sweet orange, and avoid sour orange, grapefruit, and shaddock.
2. Use tristeza-free budwood, indexed by testing on Mexican lime, if it can be obtained.
3. Control aphids in nurseries by spraying with malathion (1 oz/3 gallons).

References: 45, 47, 82.

RUB

Wind damage

Symptoms resembling those of scab are often seen on all varieties of citrus after the hurricane season.

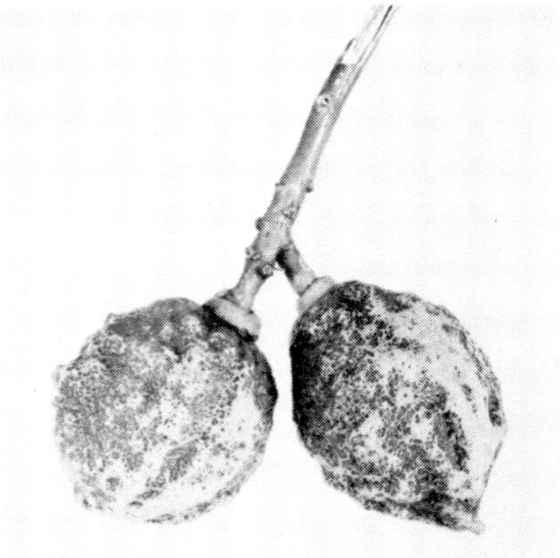

Fig. 85. Symptoms of rub on Lisbon lemons.

Symptoms
Smooth gray areas appear on young fruit and twigs. They are not raised and roughened like scab (Fig. 85).

Cause
The damage is caused when high winds whip the young fruit about, rubbing them against the bark of the large branches.

Control
1. Plant a windbreak.

References: 45.

SCAB

Elsinoe fawcettii

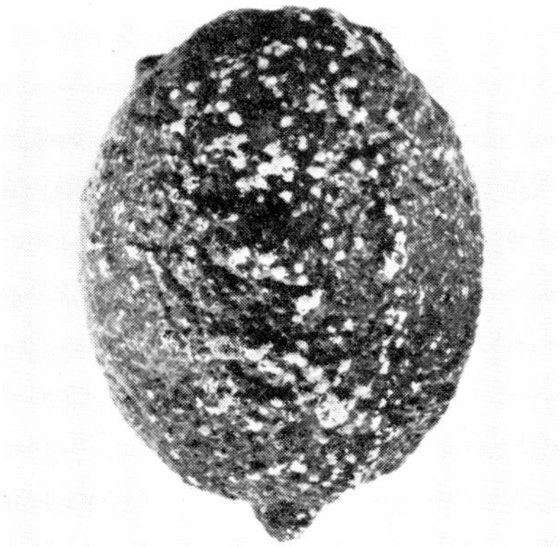

Fig. 86. Scab causing a twisting of the terminal leaves of bush lemon (above left).

Fig. 87. Scab on bush lemon twig (above right).

Fig. 88. Scab on fruit of Lisbon lemon (left).

Scab is severe only on bush lemon. It also attacks Lisbon lemon, sour orange, and occasionally seedless lime. It does not occur on kumquat, Mexican lime, and sweet orange, and it is rare on grapefruit and mandarin.

Symptoms

Raised corky scabs appear on leaves, twigs, and fruit (Figs. 86, 87, 88). Bush lemon leaves are severely distorted (Fig. 86) because the scabs form on petioles and midribs. Scabs are grey to grey-green in colour. The centres of leaf scab spots may fall out to produce shotholes. Growth of bush lemon seedlings in nurseries is retarded. Severe scab on bush lemon trees may result in dieback if they are growing under unfavourable conditions.

Disease Cycle

Only the *conidia,* produced on the scabs *(acervuli),* are locally important in spread of the disease.

In colder climates, *perithecia, asci,* and *ascospores* may be formed deep in scab tissue on dead plant parts. The *conidia* wash off the acervuli in rain and they are spread by wind. Insects may also carry them about.

Best conditions for scab development are a cool temperature (56-74°F.)

and 100% relative humidity during the night. Fruit in the post-blossom stage and leaves and twigs in a flush of growth are susceptible. More mature tissue is resistant.

Control

1. Spray seedling bush lemons with Bordeaux (3-4-50) during flushes of growth, at weekly intervals.
2. Spray Lisbon lemons and sour oranges with Cuprozin (2 oz/3 gallons) at about 20% blossom fall, and at intervals of 3 weeks for 3 applications afterwards.

References: 29, 45, 65.

SOOTY MOULD

Capnodium citri

Sooty mould is not a parasitic disease. It may retard ripening or interfere with sunlight if infestation is very high. It is associated with high infestations of sucking insects.

Symptoms
A black mould (Fig. 89) covers leaves, twigs, and fruit. The growth is entirely superficial and it can be rubbed off. The leaf area covered by the mould may be yellowed.

Fig. 89. Sooty mould with white scale insects on citrus leaf.

Disease Cycle
White flies, aphids, scale insects, and mealy bugs excrete honeydew, a sticky sweet substance that adheres to plant surfaces. The sooty mould fungus grows profusely into the honeydew but it does not infect the underlying leaf or bark.

Control
1. Control the insects involved according to the advice of the Entomologist.

References: 29, 47.

STYLAR END ROT

Disturbed water relations

This trouble was observed in 1966 in seedless limes from Suva and Seaqaqa in Fiji, following a severe drought.

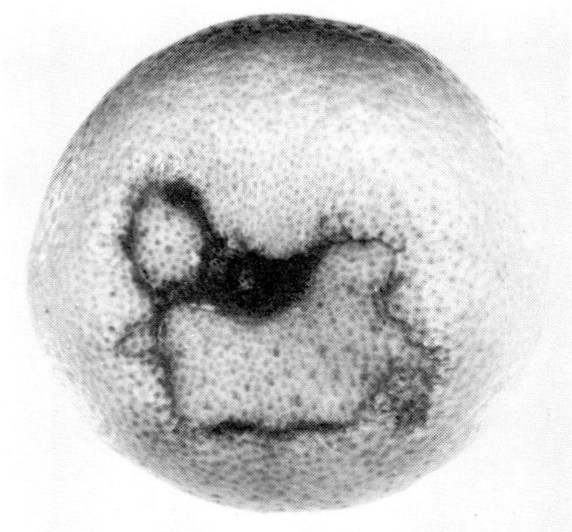

Fig. 90. Stylar end rot of seedless lime.

Symptoms

A breakdown occurs at the blossom or stylar end of the fruit (Fig. 90). The depressed area is firm, dry, and whitish. The internal tissue is dry and collapsed. The rot may extend until the lower half of the fruit is invaded. Fungi and bacterial soft-rotting organisms may invade the dead tissue and cause fruit drop.

Cause

Stylar end rot is reported elsewhere to be due to a disturbance in the water supply to developing fruit. Prolonged drought during the winter of 1965 probably caused the local trouble.

Control

1. Pick fruit while it is still green, or do not allow it to ripen on the trees.
2. Try to irrigate during drought.

References: 45, 47, 65.

Vein Enation Virus

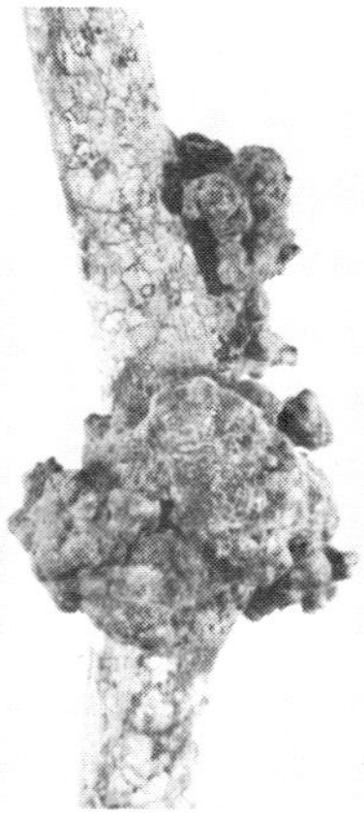
Fig. 91. Galls on bush lemon twig.

Woody gall is common on bush lemons everywhere in Fiji. The growths have little effect on the trees other than slightly reducing their vigour. Some disorders may occur when infected bush lemons are used as rootstocks.

Symptoms

Large, warty growths (Fig. 91) develop at any point on twigs or branches. They vary in size from swellings ¼-½ inch in diameter to masses about the size of a man's fist. Twigs growing at or near galls are often arranged like fans. Bark cracking occurs around the galls.

Seedling bush lemons and sometimes limes have thickened veins, caused by spindle-shaped growths called *enations*. These are miniature galls corresponding to those on the twigs and branches. They form on the lower sides of the leaves and they are matched by depressions or dimples on the upper sides.

Citrus scions grafted on to infected bush lemon stocks develop shoe-like galls at the bud union.

Disease Cycle

The virus causing woody gall was known for years as the *vein enation virus.* It was recently shown to be identical to that causing woody gall. It is transmitted locally by an aphid, *Toxoptera citricidus,* which is most abundant on citrus. This aphid also transmits tristeza virus. Some pathologists think that tissue infected by vein enation virus must be wounded in some way before woody gall will develop. The galls are outgrowths of the cambium, pushing up through the overlying bark. Aphids that transmit the disease feed only at the tender tips of shoots.

Control

1. Control aphids in nurseries by spraying frequently with malathion (1 oz/3 gallons).
2. Dc not use infected bush lemon seedlings as rootstocks.

References: 45, 82, 83.

ALGAL DISEASE

Cephaleuros pulvinatus

Algal disease is common on cocoa and coffee in Fiji and Western Samoa. It rarely affects the leaves seriously but it can cause bark cracking under certain conditions.

Fig. 92. Cracking of cocoa twigs, caused by algae.

Symptoms

Small, round, orange-yellow cushions about ¼ inch in diameter form on the upper sides of the leaves, and sometimes on the bark. The bark on young twigs cracks in a platelike fashion (Fig. 92). Dieback may follow a serious algal attack.

Disease Cycle

Algal disease does not affect vigorous cocoa. It may attack the bark of trees suffering from sudden removal of shade after a hurricane. The disease is spread by the oval *sporangia* of the alga. These are detached during windy, rainy periods and they blow to healthy leaves and bark. There they germinate in water droplets by producing *zoospores* or swimming spores similar to those of *Phytophthora.* Invasion of leaf and bark tissues is very superficial. The bark cracks after the outer tissues dry out.

Control

1. Provide adequate shade for young cocoa.
2. Apply fertilizer and black pod sprays regularly.

References: 7, 81.

ANTHRACNOSE

Glomerella phomoides

Anthracnose is common on trees growing under unfavourable conditions, and also in unshaded nurseries.

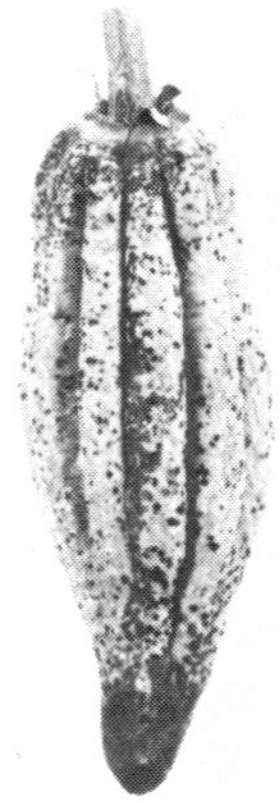

Fig. 93. Anthracnose on cherelle.

Symptoms

A large number of peppery brown spots (Fig. 93) develops on cherelles less than 3 inches long. Large brown spots may occur on older pods. After the spots run together, the pods shrivel and die.

Cocoa seedlings develop ashy grey spots about ¼ inch in diameter.

Disease Cycle

A common anthracnose fungus, *Glomerella phomoides,* can be isolated consistently from spots on pods and leaves. It attacks tissues weakened by an unfavourable environment, e.g., lack of shade, low fertility, or partial ringbarking by canker fungi. It is quite possible also that the feeding sites of aphids and other insects are invaded.

The fungus is very common on decaying fruit and leaves of all kinds of plants. It is spread by its *conidia* that are produced in great slimy masses on the surface of spots.

Control

1. Provide shade for seedlings.
2. Control black pod.
3. Apply fertilizer regularly.

References: 81.

BLACK POD AND CANKER

Phytophthora palmivora

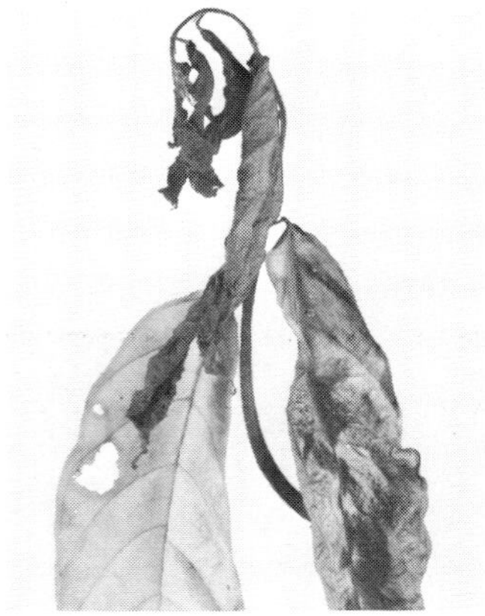

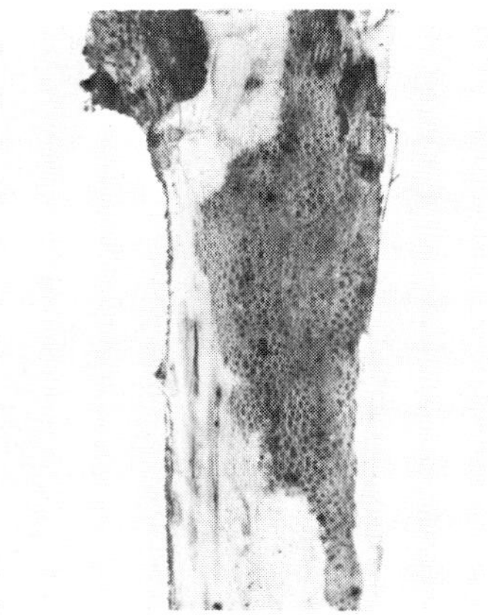

Fig. 94. Black pod (above left).

Fig. 95. Chupon blight (above right).

Fig. 96. Canker (left)

Black pod and its associated canker is unquestionably the worst disease of cocoa in the South Pacific. It causes serious reductions in the yield and it also kills trees. Amelonado and Criollo cocoa are more susceptible than the Trinitario type.

Symptoms

The most striking symptoms are the black areas (Fig. 94) that cover the pods. These spread rapidly to cover the entire surface, which then produces a white, powdery mass of spores. Then the pods dry out and the beans may rot inside. Pods of any age from cherelles less than 2 inches long to mature ones may be attacked.

Under very favourable conditions the chupons blight (Fig. 95), and often cankers develop on the branches at the base of chupons. Cankers may also commence from infected cherelles or pods. The cankers may become large enough to ringbark the trees. If the bark is cut away, the canker shows as a purplish-brown area with a black border (Fig. 96).

Cankers may also form underneath the flower cushions and infect the pods through the pedicels.

Dieback is the final stage of black pod disease after cankers have ring-barked the trees.

Disease Cycle

Sporangia produced on the surface of black pods are carried to healthy trees by wind, rain, and insects. They germinate in water droplets to produce *zoospores.* These grow *germ tubes* that can infect healthy or injured pods or cherelles. Five days after infection, the typical black spots appear and during the next 2 weeks, sporangia are produced in great numbers. The pods are then invaded by secondary fungi and boring insects. Sporangia are also produced on chupons but they are rarely found on cankers.

The fungus can also be transmitted by pruning tools used to cut infected pods and then healthy pods, or branches or chupons.

Best conditions for black pod are a high rainfall, lack of air movement, and persistent dews and mists. Cocoa planted on coastal flats or in inland river valleys with an average annual rainfall of 120-180 inches suffers more severely than plantations on higher ground, with plenty of air movement and an annual rainfall of 70-110 inches.

Control

1. Where black pod is not severe (less than 25%), cut out and burn infected pods, cherelles, and chupons regularly. Trim cankers back to healthy wood, flood with a solution of 1 part of sodium hypochlorite (Javex or Clorox) in 10 parts of water and paint with Bordeaux paste or coal tar. Sterilize pruning knives and shears with Javex or Clorox solution.
2. Where black pod is severe (more than 25%), spray the *bearing area* of trees every 2 or 3 weeks with Copper-Sandoz, Cupoxide, or Perenox, (2 oz/3 gallons) (4 lb/100 gallons) at the rate of 50 gallons per acre, depending upon the tree population. Add 2 oz of a spreader-activator such as Agral LN, X-77, or Sandovit wetting agent to every 100 gallons of spray. First carry out the sanitary measures listed in (1).

OR Spray by misting machine with Copper-Sandoz, Cupoxide, or Perenox, (1 lb/2 gallons) at the rate of 6 gallons per acre. Add 3 oz of Agral LN, X-77, or Sandovit wetting agent *first* to the water and then mix in the fungicide.

3. Do not interrupt spraying during rainy periods.

References: 7, 34, 81, 89.

BLACK SPOT

Suckling Bugs

Black spot commences with an insect injury. It can cause loss of cherelles and stunting of twigs where the sucking bug infestation is heavy.

Fig. 97. Black spot on Amelonado pod.

Symptoms

Round, purple to black spots about ¼ inch in diameter appear on the blossom-end of light-coloured pods (Fig. 97). Sometimes the spots are raised and corky. Attacked cherelles wilt. The internodes of twigs are shortened and trees show a false "witch's broom". Leaves are curled, stunted, and show star-shaped black spots and shotholes. Chupon wilt and dieback of branches may follow.

Disease Cycle

Black spot is caused by sucking bugs that inject a *toxin* or poison into their feeding sites. Then secondary fungi, e.g., *Glomerella, Botryodiplodia, Nectria, Fusarium,* invade the tissues that are killed by the toxin. If the infestation of sucking bugs is heavy, there is a repeated flushing of leaves and trees develop the appearance of a witch's broom after the bugs have fed on the buds. For some reason, certain trees become much more heavily infested than their neighbours. Sucking bugs have many hosts, including trees that are used as shade for cocoa.

Control

1. Apply Gammexane (gamma-BHC) at rates recommended by the Entomologist.

References: 81.

BROWN POD and CANKER

Botryodiplodia theobromae

Fig. 98. Brown pod with a sooty covering of spores.

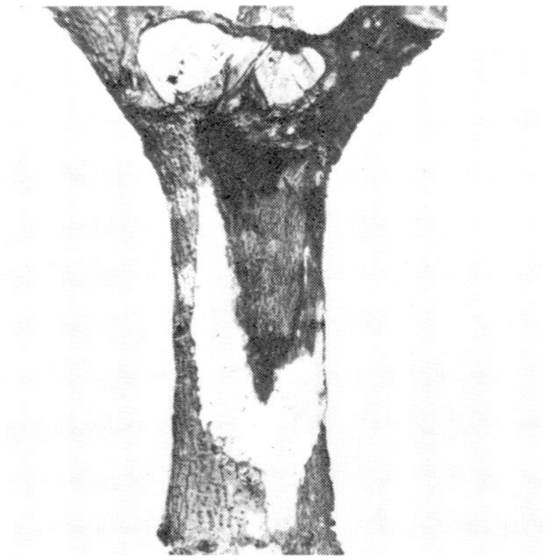

Fig. 99. Brown pod canker below severely pruned jorquette.

Brown pod and its associated canker is not nearly as serious in the South Pacific as black pod. It is found most commonly in Papua-New Guinea. It occurs on trees that are weakened by a number of environmental factors.

Symptoms

A brown spot similar to that of black pod appears on the side of the pod. It enlarges more slowly than black pod to cover the whole fruit. Brown pod is usually depressed. The pod has a roughened appearance, as if it were covered with a fine black soot (Fig. 98). Small white tendrils are also visible. Branch and trunk cankers (Fig. 99) are reddish-brown instead of purple as is the case with black pod cankers. They are also watersoaked in appearance.

Disease Cycle

The brown pod fungus is a secondary parasite with a very wide host range.

It gets into pods and bark through wounds caused by insects, man, sunburn, low fertility, and poor drainage. It cannot possibly invade healthy tissue. The fungus is spread by the dark brown *conidia* that ooze out of *pycnidia* in long tendrils during moist weather.

Control

1. Avoid poorly drained sites for planting cocoa.
2. Prune dead twigs, branches, and infected pods and cherelles.
3. Apply fertilizer regularly.
4. Supply adequate shade.

References: 34, 70, 81.

CHERELLE WILT

Nutrient/pollination Imbalance

Cherelle wilt is not a parasitic disease. It is a condition that limits the number of pods that can be set and matured on a tree under existing environmental conditions.

Fig. 100. *Left:* healthy cherelle; ***right:*** wilted cherelles.

Symptoms
At any stage up to about 3 inches length, cherelles may wilt or wither (Fig. 100). They are then invaded by secondary fungi but they remain hanging from the branches.

Cause
In the absence of black pod infection and insect injuries, the cocoa tree cannot possibly mature every cherelle that is set after a successful pollination. In this respect the disorder is similar to premature nutfall of coconuts. Incidence of cherelle wilt is higher on marginal soils and in cocoa interplanted with other tree crops. Potash may in many cases be the key nutrient.

Control
1. Apply fertilizer regularly in accordance with the results of soil tests.

References: 39, 81.

COLLAR ROT

Corticium sp.

Collar rot occured in 1967 in one nursery near Rakiraki, Viti Levu, Fiji. Only single plants were attacked.

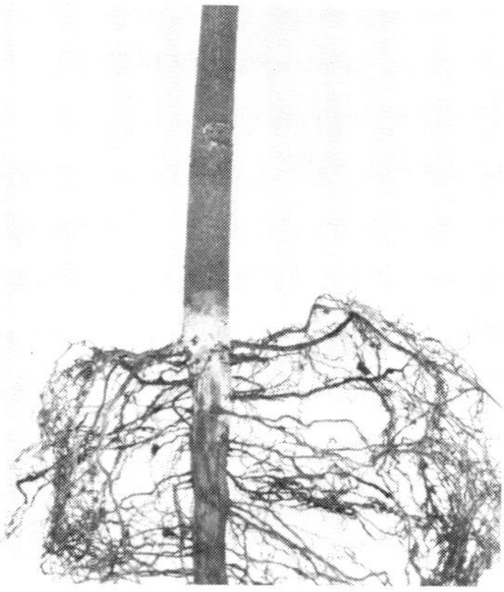
Fig. 101. Collar rot on cocoa seedling.

Symptoms
Seedlings wilt and drop their leaves. The stems at soil line may be covered by a white fungus growth (Fig. 101). The tap roots show a cracking of the bark and roots are rotted.

Disease Cycle
Collar rot is caused by a species of *Corticium,* similar to those that cause thread blight of cocoa, coffee, coconuts, and many other tropical trees. The fungus lives in undecayed organic matter that is accidentally introduced into nurseries in soil.

Control
1. Keep trash of any kind out of compost soil.
2. Drench soil with formalin solution (1 gallon of 40% concentrate/49 gallons water) at the rate of 2 pints per square foot. Cover soil for 48 hours with bags or a tarpaulin after drenching. Do not plant for 10 days.

References: 36, 81.

DIEBACK

Nonspecific

Dieback is a disease of non-specific origin that occurs everywhere in the South Pacific. It is the major cause of loss of trees of all ages.

Fig. 102. Dieback of cocoa caused by bad drainage.

Symptoms
Trees lose their leaves starting at the tips of the twigs and progressing downwards. Twigs then die back from the tips and eventually whole branches (Fig. 102) and the trees themselves die.

Cause
Dieback can be due to:

(1) *ringbarking* caused by a Phytophthora-initiated canker at or below the jorquettes.

(2) *hurricane or flood* damage that defoliates trees, breaks branches, and rots the roots.

(3) *falling coconuts* that hit the jorquettes and main branches and incite cankers in the injuries.

(4) *bad drainage* that encourages root rot by various fungi.

(5) *prolonged drought.*

(6) *lack of shade.*

(7) *soil deficiencies.*

Control

1. Control black pod.
2. Prune and fertilize trees showing the first symptoms of dieback.
3. Choose well-drained planting sites and supply adequate shade.
4. Trim cankers and treat with a wound protectant.

References: 34, 81.

MARGINAL SCORCH

Potash Deficiency

Marginal scorch is common on cocoa in the South Pacific, particularly on highly alkaline old reef soils. It is also found on highly acid soils. Often marginal scorch is associated with dieback.

Fig. 103. Old cocoa leaf with marginal scorch.

Symptoms
Older leaves show a pronounced marginal scorch (Fig. 103). There may also be pale yellow areas in the interveinal zones at the edges of the leaves. Dieback and excessive cherelle wilt are often symptoms that are associated with marginal scorch.

Cause
It is generally attributed to a deficiency of potash, which becomes unavailable on alkaline soils at a soil pH of 7.5-8.5, and on acid soils at a pH below 5.5. It may also be excessive in cocoa interplanted with coconuts, which compete heavily for potash. Some marginal scorch may be caused by salt spray injury.

Control
1. Check soil fertility and apply a high-potash fertilizer in accordance with the results of soil analyses.

References: 81.

"PINEAPPLE" DISEASE

Unknown

This deformity of pods sometimes occurs towards the end of the cool season in Fiji. It was first noted during 1965 at Nadurulouiou and Koronivia.

Symptoms
Pods show a complete or partial segmentation into 4- or 5-sided bumps that resemble pineapple fruitlets (Fig.104). Pedicels may show severe bark cracking.

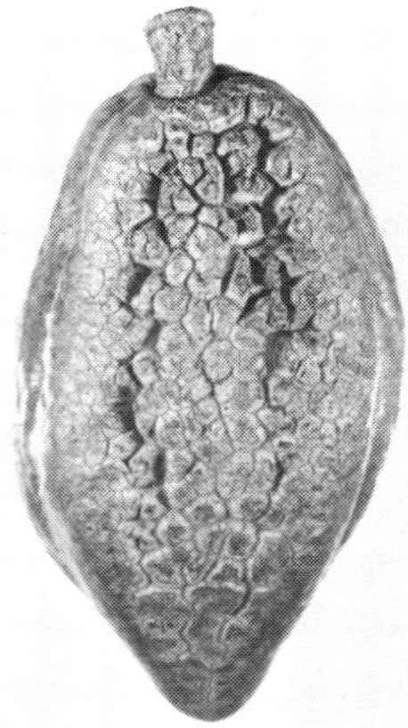

Fig. 104. "Pineapple" disease on Amelonado pod.

Cause
The cause is unknown. The red rust alga, *Cephaleuros pulvinatus,* is sometimes found associated with the condition. When wet warm weather arrives, pineapple disease fades out.

Control
No special measures are recommended.

THREAD BLIGHT

Corticium sp.

Thread blight is seen occasionally in old unpruned cocoa growing under excessive shade.

Fig. 105. Thread blight on twig and leaf of cocoa.

Symptoms

Leaves hang from the twigs. If they are examined closely they are found to be suspended by white fungus threads (Fig. 105), that extend from the bark of the twigs and fan out over the leaf surface. Leaves are blackened, withered, and sometimes completely skeletonized and reduced to a papery framework.

Disease Cycle

The fungus associated with thread blight lives on dead wood and bark in older trees. It can overgrow healthy bark and leaf tissue, which it then kills. It is spread from place to place by light spores *(basidiospores)* that are produced on the white mycelium. The diease is definitely worst under excessive shade and high rainfall.

Control

1. Prune out diseased wood and reduce shade.
2. Apply a clean-up spray of Cuprozin, Perenox, Cupoxide, or Copper-Sandoz (2 oz/3 gallons).

References: 56, 81.

BLEEDING DISEASE

Ceratocystis paradoxa

This disorder is not, strictly speaking, a parasitic disease, but a condition that develops after mechanical injury. It is very common wherever coconuts are grown.

Symptoms
A reddish-brown sap exudes from cracks (Fig. 106) in the lower part of the trunk. The underlying tissues are affected by a brown, watery rot that may ring-bark the palm.

Fig. 106. Bleeding disease of coconut.

Disease Cycle
The causal fungus, which is very common on bark, gets in to the trunk through growth cracks or wounds of any kind, e.g., the bore holes of insects, burns, climbing cuts, etc. It is the same fungus that is associated with "pineapple" disease of sugarcane setts and butt rot of pineapples. It is a wound parasite that can invade only weakened or injured tissues.

Control
1. Discourage people from cutting steps into the bases of palms, particularly when they are young.
2. Cut out affected parts and dress the wounds by painting with Bordeaux or coal tar.

References: 12.

BUD ROT

Phytophthora palmivora

Fig. 107. Early stage of coconut bud rot (above).

Fig. 108. Coconut crown decayed by bud rot (left).

Bud rot has been observed only occasionally in Fiji on the island of Taveuni. The Malayan Dwarf variety, which is sometimes interplanted with the local tall variety, is attacked under favourable conditions.

Unconfirmed reports of bud rot have been received from the island of Eua in the Tonga group and from the New Hebrides. Its presence has been confirmed in Papua–New Guinea. In the Philippine Islands it is sufficiently serious to require special control measures.

Symptoms

The *youngest* leaves of palms wilt and collapse. Leaves die in succession from the centre of the crown (Fig. 107). The bud is often found to be a stinking rotten

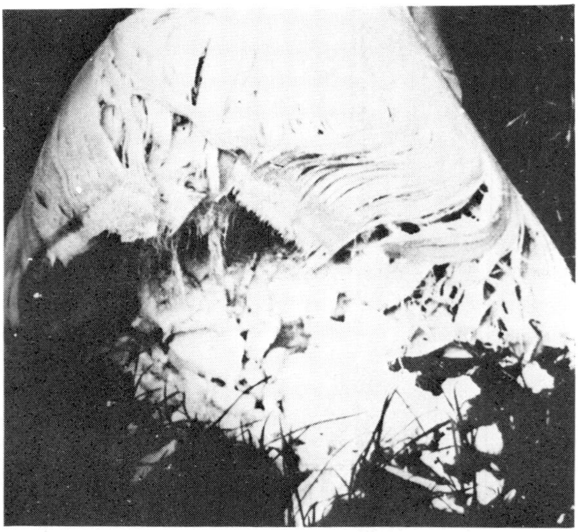

Fig. 109. Close-up of advanced stage of bud rot.

mess (Fig. 109). The bases of young leaves may be affected by brown, sunken spots (Fig. 108). Palms do not usually recover from the disease.

Disease Cycle

The bud rot fungus is the same as the one that causes black pod of cocoa, which may in some cases supply spores for palm infections. *Sporangia* are produced on infected coconuts or cocoa. During rain the sporangia are blown about and they land on healthy palms, where they germinate to produce *zoospores.* These spores infect the young leaf bases and the resulting *mycelium* then invades the buds. Secondary soft-rotting bacteria then finish the job.

The best conditions for infection are a relative humidity of 100% and temperatures in the 70-80°F. range. These conditions are more likely to be found about 500-600 feet above sea level. Daily mists also contribute to a high rate of infection.

Control

1. Cut out the diseased palms as soon as they are detected.
2. Improve the circulation of air by cutting out growth of shrubs and weeds under palms.
3. Avoid planting Malayan Dwarfs under the cover of tall varieties, in areas favourable for development of bud rot.

References: 22, 56, 70.

GREY LEAF SPOT AND BLIGHT

Pestalotiopsis palmarum

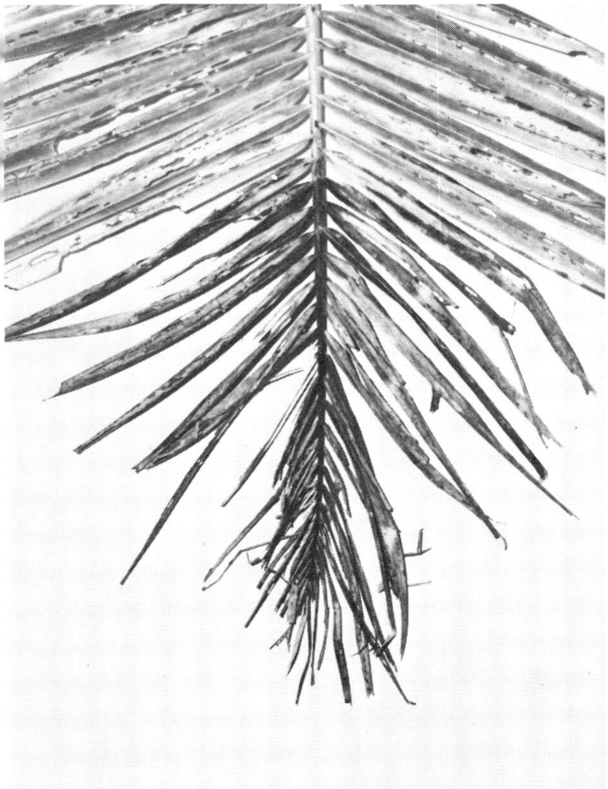

Fig. 110. Coconut leaf blight.

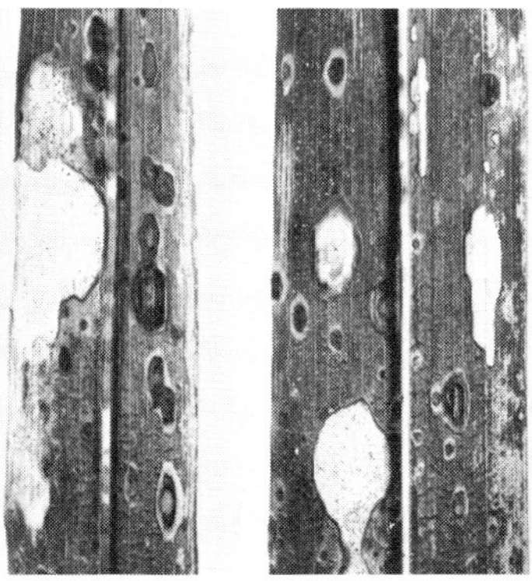

Fig. 111. Early stage of grey leaf spot.

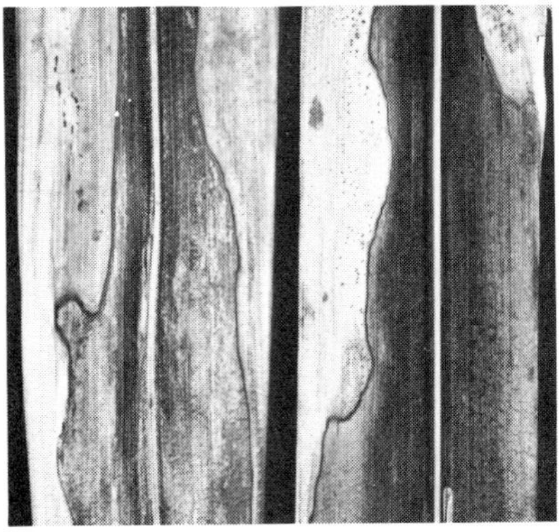

Fig. 112. Advanced stage of grey leaf spot.

This is a very common disease of coconuts in the South Pacific. It seems to affect the Malayan Dwarfs more severely than the standard talls. The yellow Malayan Dwarf is quite susceptible. It never causes defoliation.

Symptoms
Small, yellow angular spots with grey bands develop on the older leaves. The centres of the spots later turn ashy-grey while the outside bands turn dark brown (Fig. 111). Spots may run together to blight the tips and edges of fronds (Fig. 110). Spots are covered with black pinpoints (Fig. 112). In severe cases there is leaf drop, but never complete defoliation.

Disease Cycle
The black pinpoints on the leaves are fruiting bodies *(acervuli)* that contain masses of *conidia.* These have fishhook-like appendages that may aid in dispersal. Conidia infect the leaves probably when they are quite young. The spots do not develop until the leaves begin to age. Malayan Dwarfs appear to be more susceptible probably because their life span is shorter than that of standard talls.

Palms growing in poor soil show more blight than those in deep, fertile soil.

Control
1. Cut out and burn severely diseased leaves.
2. Apply a complete fertilizer high in potash, according to the recommendations of the Soil Chemist.

References: 12, 22, 39, 70.

PREMATURE NUTFALL

Nutrient/pollination imbalance, Rat and hurricane damage, Drought

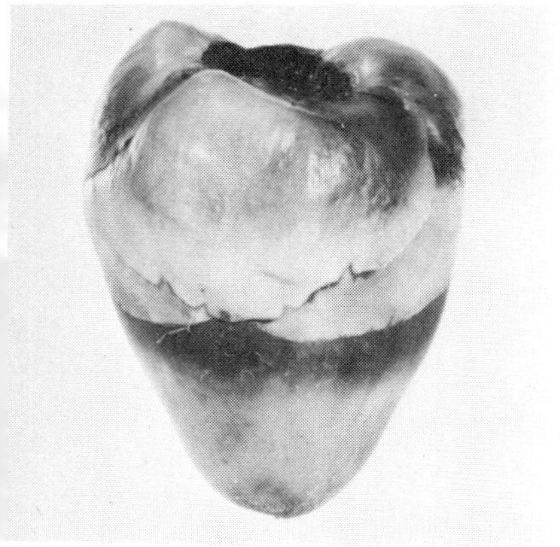

Fig. 113. "Button" nut found on the ground.

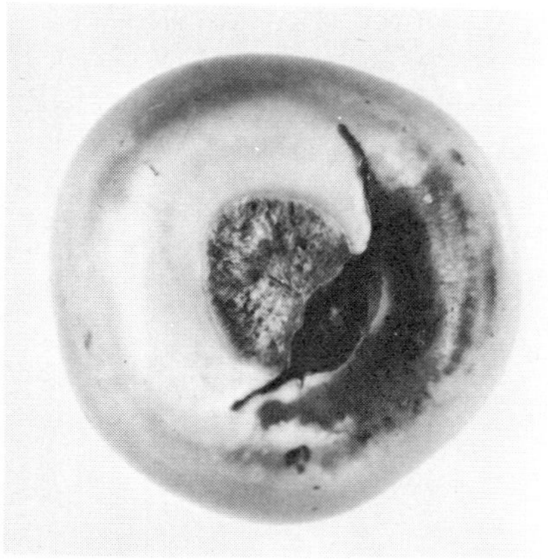

Fig. 114. Young nut injured by strong wind and invaded by fungus.

All coconuts experience premature nutfall to varying degrees. It is more evident in palms growing under unfavourable conditions.

Symptoms

Nuts fall in the "button" stage (Fig. 113) within 2 months of flowering, or at any stage up to 4-5 months after flowering. The size of nuts shed at the later stage varies from 1½-3 inches in diameter. Chewing marks and rotting (Fig. 114) are evident on the abscission layers.

Cause

"Button" nutfall is due to the fact that palms cannot possibly mature every nut they set. Most of the nutfall at this stage may be normal, depending upon soil fertility and rainfall. It is a form of self-pruning.

Premature nutfall after 2 months may be due to drought, rat injury, delayed hurricane damage, and fungal invasion of the pedicels and abscission layers of damaged nuts.

Control

1. Correct soil deficiencies.
2. Control rats.

References: 12.

SEEDLING BLIGHT

Helminthosporium incurvatum, Ceratocystis paradoxa, Various fungi

Most seedling coconuts suffer from blight before they become established. It is common on poor soils.

Fig. 115. Blight on 6-months seedling.

Symptoms

Seedlings are yellowed and covered by greyish-brown spots of irregular size (Fig. 115). Sometimes a greenish-brown powdery mould may be seen on the spots. Tips and edges of the leaves may blight so severely that only the midribs are left intact. Seedlings usually recover.

Disease Cycle

The fungi that cause seedling blight may not be limited to coconuts. Some of them can be found on almost any kind of dead or decaying organic matter, or in soil. They attack weakened seedlings growing in poor soils or in partial shad If only nitrogen is applied as a remedial measure, the condition becomes worse

Control

1. Locate coconut nurseries in good soil in full sunlight.
2. Apply a complete fertilizer in severe cases, as recommended by the Soil Chemist.

References: 12, 22.

Bad Drainage

Fig. 116. Tapering and reduced crown of coconut in poorly drained land.

Fig. 117. "Pencil point" rhino beetle injury in tide land.

In Fiji this is very common between Yaqara and Nadi on the King's Road. It can be mistaken for much more serious disease, such as cadang cadang, which is not known to exist in the South Pacific.

Symptoms

Leaves are reduced in size, yellowed, but not to the same extent as in yellows encountered on land low in fertility. The trunk tapers towards the top (Fig. 116), and when the crown finally dies and falls off, it resembles a pencil point (Fig. 117).

Cause

The disorder is very common on poorly drained soil. The coconut does not do well in or near mangrove swamp or in cane fields close to tide land. Hardpan soils, very sandy soils, and senility may also produce it. The rhino beetle undoubtedly hastens death of the crown.

Control

1. Choose well-drained sites for planting.

References: 12, 39.

THREAD BLIGHT

Corticium penicillatum

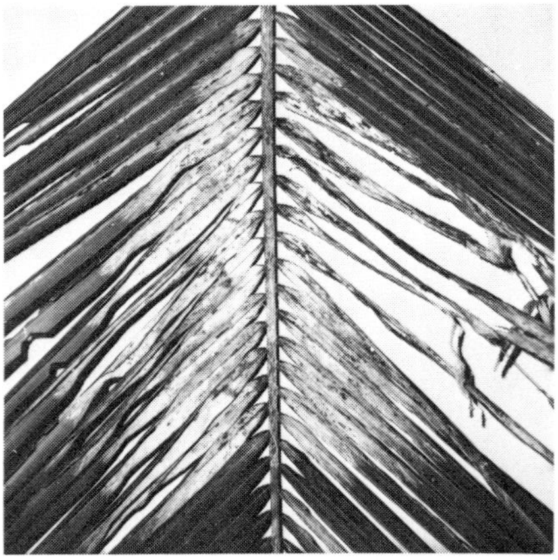

Fig. 118. Thread blight patch on underside of coconut leaf.

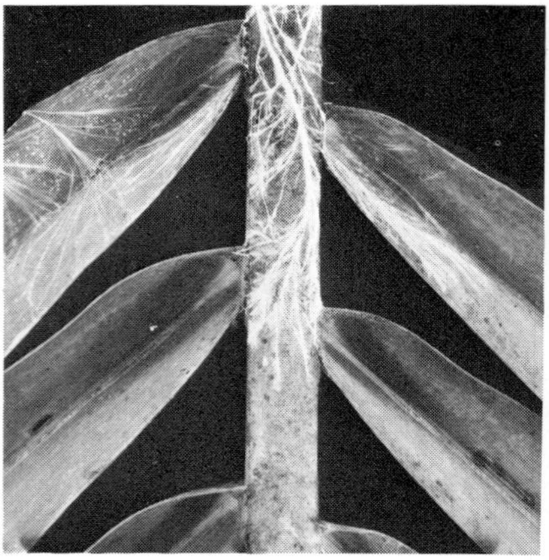

Fig. 119. Close-up of thread blight fungus on underside of a leaf.

Thread blight occurs in inland river valleys but it is rarely seen along coast lines. It is not a serious disease.

Symptoms

Leaves turn grey and brown in large patches anywhere on their surfaces (Fig. 118). From a distance thread blight resembles grey leaf spot very closely but the latter disease attacks the tips and margins of the leaves. White strands of fungus growth can be seen on the undersides of the leaves (Fig. 119), along the midribs and out along the leaflets. In severe cases leaves may fall off.

Disease Cycle

Under the extremely wet and shaded conditions under which coconut palms are growing in certain inland river valleys, the thread blight fungus gets a start on dead leaves and other organic matter. Spores are produced on the felty growth and they are blown to healthy palms. Leaf tissues are killed by the fungus as it advances.

Control

1. Avoid very wet and shaded zones for coconut production.
2. Eliminate excessive shade or use a wider spacing.
3. Cut out and burn infected leaves.
4. Spray the remaining leaves with Kopi, Cuprox, or Perenox (2 oz/3 gallons).

References: 22, 70, 71.

TWISTING

Hurricane

Twisting of the crowns of coconuts is often seen and it causes considerable concern.

Symptoms
Fronds are twisted to one side of the crown (Fig. 120), often through an angle of more than 180°. Leaf breaking, drooping, and dropping may occur.

Fig. 120. Severe twisting of a coconut palm in Vatulele island, Fiji.

Cause
The trouble is caused by steady winds blowing with hurricane force for a prolonged period from one quarter. The injured tissues are attacked by secondary rotting fungi such as *Botryodiplodia, Ceratocystis,* and *Fusarium.*

Dying of leaves takes place often too late for people to remember that a hurricane has occured. The original damage may be accentuated by insect attacks and drought.

Control
1. Cut out and burn broken fronds.
2. Apply fertilizer to rehabilitate palms if the cost is economically feasible.

BROWN EYE SPOT and BERRY BLOTCH

Cercospora coffeicola

Fig. 121. Eye spot on leaf of Robusta coffee.

Fig. 122. Berry blotch of Arabica coffee.

Brown eye spot causes defoliation of seedlings in nurseries but it is not important on mature bearing plants. Berry blotch causes some loss of fruit of both Arabica and Robusta.

Symptoms

Spots that vary from $1/8$-$1/3$ inch in diameter, with grey centres, brown edges, and yellow haloes appear on the leaves (Fig. 121). Leaves are killed and drop off if the attack is severe.

Berries develop large, dark grey or brown spots (Fig. 122) covered with a felty green growth of fungus. They turn black, shrink, and fall off before harvest.

Disease Cycle

Very little is known about the life cycle of the eye spot fungus and its ability to attack plants other than coffee. It is spread by the sickle-shaped *conidia* that are produced in great numbers, particularly on berries.

Best conditions for the disease appear to be high rainfall and lack of shade. Neglected old coffee and very young seedlings are most severely attacked. Arabica seems to be more susceptible than Robusta.

Control

1. Spray nurseries with Kopi, Cuprox, or Vitagran (2 oz/3 gallons).
2. Provide adequate shade and fertilizer.

References: 13, 36, 87.

Unknown

Little leaf has been found in coffee at Koronivia, Naduruloulou, and Nacocolevu in Fiji. It is impossible to estimate its effect on yield.

Symptoms
The terminal leaves are very small, narrow, and twisted (Fig. 123). The interveinal tissues are yellow or white. Some dieback may occur but the bushes usually recover.

Fig. 123. Little leaf of Robusta coffee.

Cause
It is possible that little leaf is due to a zinc deficiency, because other crops, e.g., citrus, pineapples, maize, show such symptoms when they are grown on the same soil type.

Control
1. Use fertilizer with a trace element complement.
2. Try spraying the plants with a 0.5% solution of zinc sulphate, during flush periods.

References: 16, 87.

Hemileia vastatrix

Fig. 124. Symptoms of rust on upper and lower sides of Arabica coffee leaves.

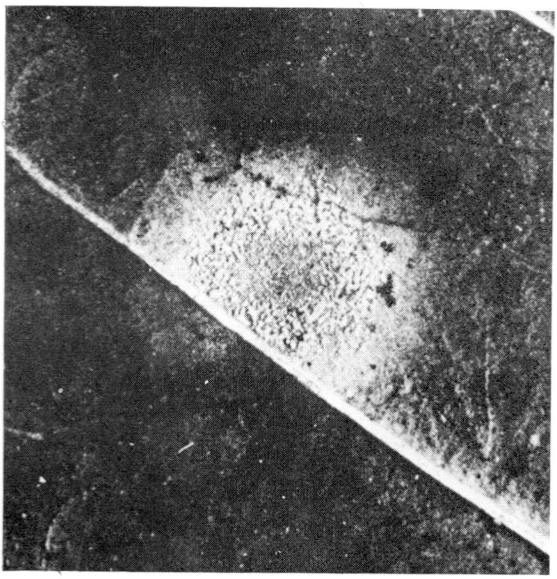

Fig. 125. Close-up of rust pustule on lower side of Arabica coffee leaf.

Rust is a serious disease on Arabica coffee planted at sea level in locations where the average annual rainfall exceeds 90 inches. It has been reported from Fiji, Samoa, New Caledonia, and the New Hebrides. Robusta coffee is rust resistant. An outbreak in Papua-New Guinea was successfully eradicated.

Symptoms

Pale yellow spots, changing to brown, appear on the upper and lower sides of leaves (Fig. 124). They vary in size from $^1/_8$ inch in diameter to large blotches.

The undersides of the spots are covered by an orange-coloured dust (Fig. 125). Leaves eventually drop and dieback follows.

Disease Cycle

The rust fungus is spread by its orange-coloured *uredospores.* Rain, wind, and possibly insects carry them about. The spores germinate in water droplets and penetrate the stomata on the undersides of leaves.

Rust develops best on crowded, unpruned coffee grown in high rainfall belts. Under good shade at high altitudes, the disease is mild.

Control

1. Grow rust-resistant Robusta varieties.
2. Space coffee bushes at the recommended distances, prune and fertilize regularly.
3. Spray susceptible varieties at 3-week intervals during the wet season with Bordeaux 3-4-50. at the rate of 70-80 gallons per acre. Direct the spray *at the undersides* of the leaves.

OR

Spray with Perenox, Cupoxide, or Copper-Sandoz (2 oz/3 gallons; 4 lb/100 gallons) at the same rate per acre. Add 2 oz spreader-activator such as Agral LN, or Sandovit wetting agent, or X-77, to every 100 gallons of spray.

OR

Spray by misting machine with Perenox, Cupoxide, or Copper-Sandoz (1 lb/2 gallons) at the rate of 12 gallons per acre. Add 3 oz of Agral LN, or Sandovit wetting agent, or X-77, to the water *before* mixing in the fungicide.

References: 16, 36, 56, 87.

SUCKER BLIGHT

Xyleborus sp. *Botryodiplodia theobromae, Glomerella phomoides, Fusarium* sp.

Suckers of both Arabica and Robusta coffee are prone to blight, especially where shade is inadequate.

Symptoms
Suckers and sometimes small branches show black areas extending for several inches (Fig. 126). Only the smooth bark is attacked. Leaves are killed and dieback follows.

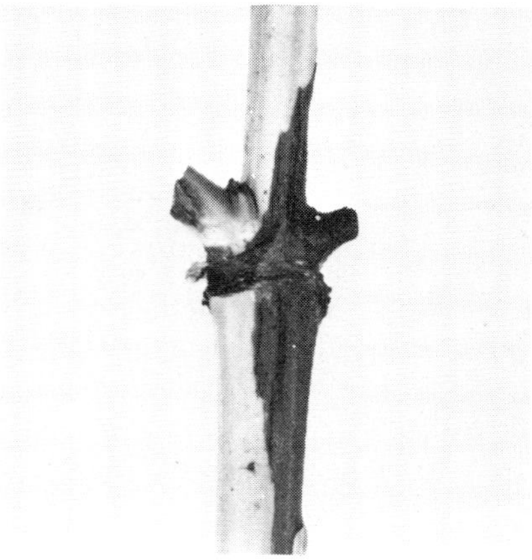

Fig. 126. Robusta coffee sucker attacked by a boring insect and blight.

Disease Cycle
It is supposed that some shock to the plant, e.g., removal of shade and a subsequent lack of moisture supply, makes the suckers susceptible to borer attack. Secondary fungus invaders, e.g., *Botryodilodia, Glomerella,* and *Fusarium,* then grow into the insect wounds and kill the wood.

Control
1. Provide adequate shade.
2. Desucker and prune and fertilize regularly.

THREAD BLIGHT

Corticium Koleroga

Thread blight has been observed in Fiji, Western Samoa, and New Caledonia and it is probably found everywhere coffee is grown in the South Pacific. It can cause loss of leaves and dieback.

Fig. 127. Thread blight of Robusta coffee.

Symptoms

Dead leaves hang by brown threads from the branches (Fig. 127). A silvery weblike growth can be found on the undersides of the leaves. Leaf tissues underneath the growth are killed. Dieback may follow leaf drop.

Disease Cycle

The cycle is identical to that of thread blight of cocoa, coconut, passion fruit, pepper, and other plants.

Control

Use the same measure as for control of thread blight of other crops.

BLOSSOM END ROT

Water and Fertilizer Imbalance

Blossom end rot of watermelons occurs on poor soils. Often it follows severe loss of leaves due to one or more fungus disease.

Symptoms
The blossom end of the fruit caves in and turns black (Fig. 128) at any time after the blossom is shed. Sometimes a soft or hard rot develops. Fruit fails to mature and it may be deformed.

Fig. 128. Young melon deformed by blossom end rot.

Cause
When a drought follows heavy rain, particularly on light soils with a low moisture-holding capacity, blossom end rot is very likely to develop. It may also be associated with a fertilizer imbalance, such as an excess of N and not enough K. It often follows defoliation by powdery mildew, downy mildew, or gummy stem blight.

Control
1. Avoid very light soils for commercial plantings.
2. Check soil fertility.
3. Control foliage diseases.

Pseudoperonospora cubensis

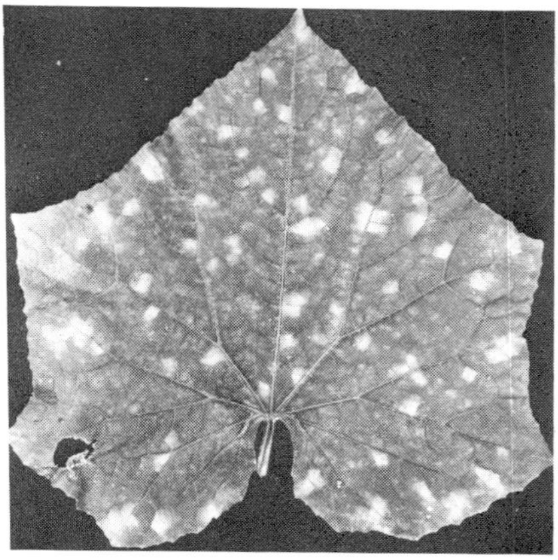

Fig. 129. Early symptoms of downy mildew on cucumber.

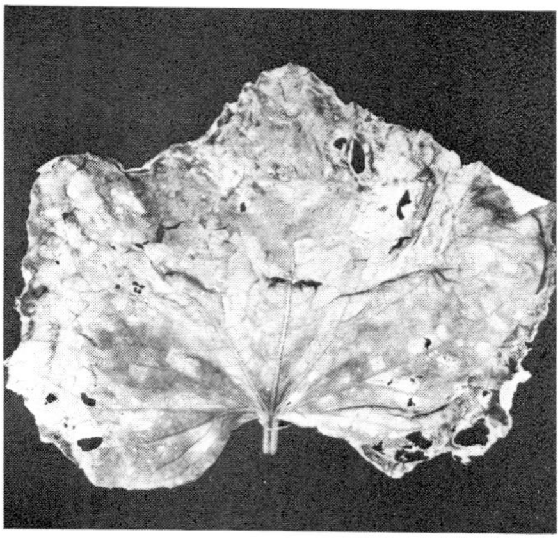

Fig. 130. Advanced symptoms of downy mildew on cucumber.

Downy mildew is severe on cucumber, rockmelon, pumpkin, squash and mild on watermelon. It occurs in both the wet and dry zones of Fiji, and also in Guam, Papua-New Guinea, and Tonga.

Symptoms

On cucumber and rockmelon, the disease begins as small, angular, yellowish-white spots (Fig. 129). On watermelon, pumpkin, and squash, the interveinal tissues become quite yellow and resemble magnesium deficiency. During favourable weather, the undersides of the leaves are covered with a purplish mould. Leaves eventually shrivel and die (Fig. 130), taking somewhat longer than in the case of powdery mildew infection. Plants are stunted and fruits are small and poorly flavoured.

Disease Cycle

The downy mildew fungus spreads by the purplish *sporangia* produced at the tips of *sporangiophores* on the undersides of the leaves. The sporangia are carried by wind to healthy leaves, where they germinate in water droplets to produce *zoospores* These swim about for a while and then form germ tubes that infect the leaves through the stomata. Within 4-5 days, a new crop of sporangia matures. Another stage in the life cycle, *oospores,* has been reported from colder countries.

Best conditions for infection and spread of disease are wet weather and temperatures between 60° and 70°F.

Control

1. Rotate cucurbits with crops belonging to families not susceptible to cucurbit downy mildew, e.g., tomatoes, peanuts, rice, sorghum, maize.

2. Commencing when the vines are about 5 feet long, spray with mancozeb (Dithane M-45) (1 oz/3 gallons) every 7-10 days as for control of gummy stem blight and Cercospora leaf spot. Apply the spray to both sides of the leaves.
3. Experiment with resistant varieties: Ashley and Polaris cucumbers; Georgia 47 and Homegarden rockmelons; Fortuna pumpkin.

References: 14, 88.

GUMMY STEM BLIGHT

Mycosphaerella melonis

Fig. 131. Leaf blight of Charleston Gray watermelon.

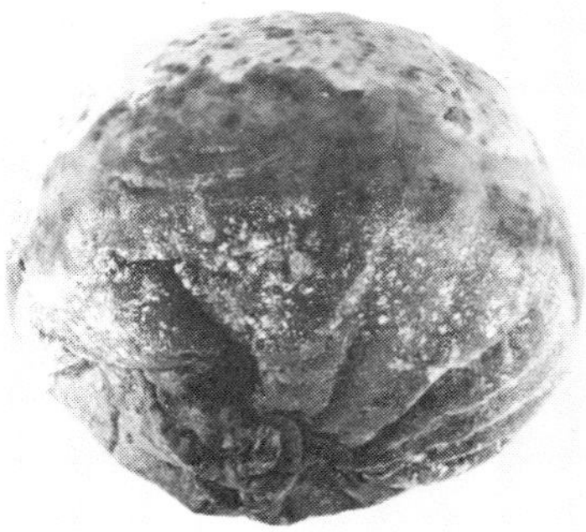

Fig. 132. Black rot on Charleston Gray watermelon.

Gummy blight is the most serious disease of watermelons in Fiji and Tonga. In Fiji it is so bad that the crop cannot be grown economically in the wet zones. It occurs to a lesser extent on rockmelon, pumpkin, and squash.

Symptoms

Starting at the centres of the plants, leaves develop irregular dark brown or black blotches (Fig. 131). Stems show pale brown streaks that enlarge, lengthen, split, and exude a black gum. Watermelon fruit shows purplish pimples sometimes with white centres and black pinpoints. A secondary soft rot (Fig. 132) may develop on fruit that is severely attacked, even during transit to an overseas market. If the crop is defoliated, fruit may show blossom end rot.

Disease Cycle

The fungus produces *conidia* in *pycnidia,* and *ascospores,* in *asci* in *perithecia.* Both *pycnidia* and *perithecia* are visible as black pinpoints on leaves and fruit. When leaves are wet, the *conidia* exude in slimy masses and the *ascospores* are ejected. The latter spores are responsible for spread over long distances. The fungus is also seed and soilborne. Infection takes place through stomata.

Best conditions for gummy stem blight are wet weather and a temperature around 80°F. Secondary soft rot of fruit is worst when wet weather persists during harvest.

Control

1. Plant cucurbits for export only in the dry season and in zones with an annual rainfall of 70 inches or less.
2. Do not plant seed until well into the dry season.
3 Rotate cucurbits with non-susceptible crops outside the cucurbit family.
4. Spray with mancozeb (Dithane M-45) as for control of downy mildew and Cercospora leaf spot.

References: 14, 36, 88.

LEAF SPOT

Cercospora citrullina

Leaf spot is a minor disease of cucurbits everywhere in the South Pacific. It is serious only where soil fertility is very low. It has been observed on all hosts, but is more obvious on watermelon, kerala *(Momordica charantia),* cucumber, and rockmelon, than on pumpkin and squash. It is often mistaken for gummy stem blight, a more serious disease.

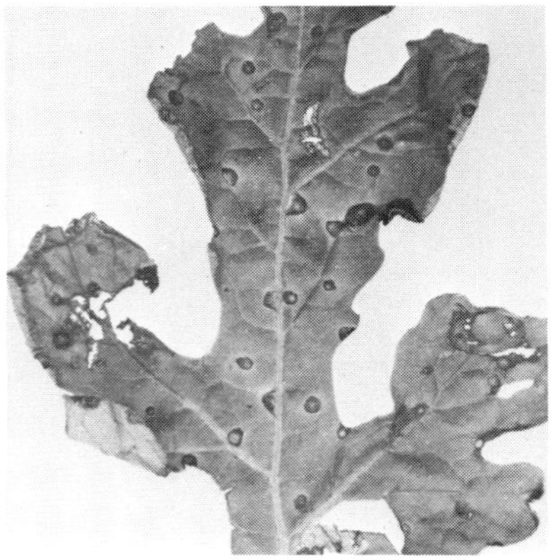

Fig. 133. Round spots of *Cercospora* on watermelon.

Symptoms

On watermelon and kerala the spots are circular, pale brown or white, with black edges (Fig. 133). They are about ⅛ inch in diameter. On cucumber and rockmelon, the spots are angular and about ¼ inch across, they may be distinguished from downy mildew spots by the presence of greenish-brown masses of spores, and by the absence of a purplish mould on the undersides of the leaves. Leaf spot rarely causes severe loss of leaves.

Disease Cycle

The pathogen is *Cercospora citrullina.* Like many other species in this group, it may be able to infect and live on plants other than cucurbits. The fungus is spread by sickle-shaped spores produced on the spots on both sides of the leaves, but mostly on the upper surfaces. New infections occur every 7-10 days during wet, warm weather.

Control

1. Spray with mancozeb (Dithane M-45) as for the control of downy mildew and gummy stem blight.
2. Rotate cucurbits with non-susceptible crops, e.g., maize, sorghum, rice, tapioca, on a 3-year basis.
3. Check soil fertility.

References: 14.

Cucumber Mosaic Virus

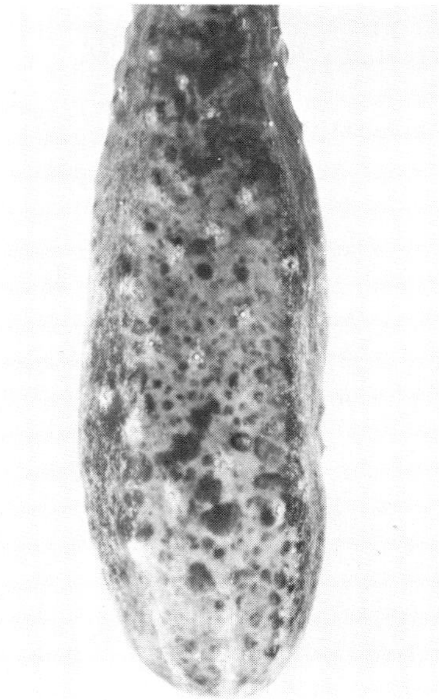

Fig. 134. Mosaic on cucumber leaf (above left).

Fig. 135. Mosaic on cucumber fruit (above right).

Fig. 136. Mosaic on watermelon leaf (left).

Mosaic has been observed on cucumber, rockmelon, squash, pumpkin, and watermelon. It is most serious on cucumber and rockmelon, because the fruit from infected vines is useless. Squash and pumpkin are relatively tolerant. The disease appears to be building up in watermelon.

Symptoms

Leaves of cucumber and rockmelon are mottled (Fig. 134), distorted, stunted, and curl downwards at the edges. Vines are abnormally short. The leaves in the centres of the plants turn yellow and die. Cucumber fruit is characteristically warted and covered with pale and dark green areas (Fig. 135).

Pumpkin and squash are less severely affected, usually showing a mild mottling on the leaves. The leaves of watermelon show a mottling (Fig.136), and they have a tendency to cup upwards and are smaller than normal. Vines are stunted and yield is reduced.

Disease Cycle

The cucumber mosaic virus, of which there are many strains, is transmitted by *Aphis gossypii,* which is found in great numbers on cucurbits and other hosts. It can also be transmitted by handling infected and healthy plants, and by machinery. It infects many weeds. A very low percentage of cucurbit seeds may carry the virus.

The strain of mosaic on watermelon may be distinct from that on cucumber and rockmelon.

Control

1. Obtain seed from virus-free sources.
2. Rogue infected plants and burn.
3. Control weeds.
4. When aphids appear, spray with malathion (1 oz/3 gallons). It can be included in the maneb spray for control of fungus diseases.
5. *DO NOT USE DDT ON CUCURBITS:*
   It does not control aphids and it causes serious injury.
6. Try mosaic-resistant varieties, e.g. Shamrock, Ilima, Ohio MR-200 cucumbers.

References: 11, 14, 74, 88.

POWDERY MILDEW

Erysiphe cichoracearum

Powdery mildew is severe on all cucurbits throughout the South Pacific. Watermelon, cucumber, rockmelon, pumpkin, and squash are all susceptible. In some years, it can kill the entire crop during the cool, dry season.

Fig. 137. Powdery mildew on cucumber leaf.

Symptoms

A white powdery covering (Fig. 137) develops on both sides of the leaves, which then shrivel and die. Mildew produces raised pimples on watermelon fruit.

Disease Cycle

The powdery growth on the leaves consists of chains of barrel-shaped spores that are detached from the leaf surface and that blow about to spread the disease to healthy leaves. The spores *(conidia)* are capable of germinating *without* water droplets, in a relative humidity a little below 100%. The new growth of mildew is evident in 8-10 days. In tropical areas, the fungus probably lives continuously in the conidial stage, without producing resistant bodies *(perithecia)*.

Conditions for disease development are best during the cool, dry season, when air temperatures are between 68° and 75°F., and when the soil is dry.

Control

1. Plough diseased crop refuse to a depth of 9-10 inches.
2. Keep midew-susceptible weeds under control.
3. Include karathane (1 oz/3 gallons) in 2 or 3 of the mancozeb sprays, applied for control of other foliage diseases, *after* mildew appears.
4. Experiment with resistant varieties: Ashley and Polaris cucumbers; PMR-45, Georgia 47, Homegarden rockmelons.

References: 14, 36, 69, 88.

YELLOWS

Deficiency

Yellows has been observed on cucurbits on alluvial soils in the Rewa Valley of Fiji. Pronounced symptoms were evident on luffa *(Luffa cylindrica).*

Fig. 138. Symptoms of yellows on leaf of luffa *(Luffa cylindrica).*

Symptoms

On the older leaves the interveinal tissues turn almost white (Fig. 138). The veins remain banded by dark green. There is no evidence of downy mildew infection, as with interveinal yellowing of watermelon. Plants do not appear to suffer otherwise.

Cause

The exact cause is unknown. Symptoms closely resemble descriptions of zinc deficiency in cucurbits. However, the leaves are not reduced in size, as is often the case where this element is lacking. The soil concerned is known to be highly acid (pH 4.5). In the same location, maize showed symptoms of white leaf, and citrus of little leaf, both of which may be attributed to zinc deficiency.

Control

1. Check soil fertility and acidity.
2. Apply fertilizer and lime in accordance with the recommendations of the Soil Chemist.

References: 14.

BROWN LEAF SPOT

Cladosporium colocasiae

Most dalo varieties are affected but the disease causes no economic losses.

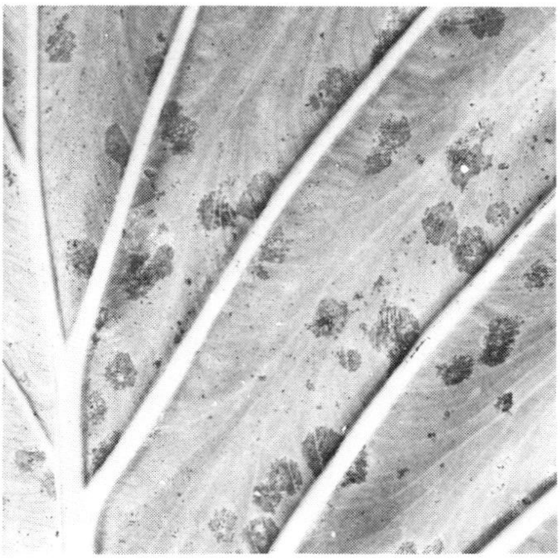

Fig. 139. Brown leaf spot on the variety Samoa.

Symptoms
Round, reddish-brown blotches (Fig. 139) about $\frac{1}{4}$ - $\frac{1}{2}$ inch in diameter develop on both sides of the leaves. They are entirely superficial.

Disease Cycle
The pathogen may be a weak parasite that gets its start in the secretions of aphids, white flies, and scale insects that are common on dalo. *Conidia,* produced on long, brown *conidiophores,* are responsible for spreading the disease. Very little is known about the life cycle of the fungus or the conditions that favour it.

Control
No special measures are recommended.

References: 36, 56.

CORM ROT

Phytophthora cinnamomi Phytophthora nicotianae var. *parasitica*

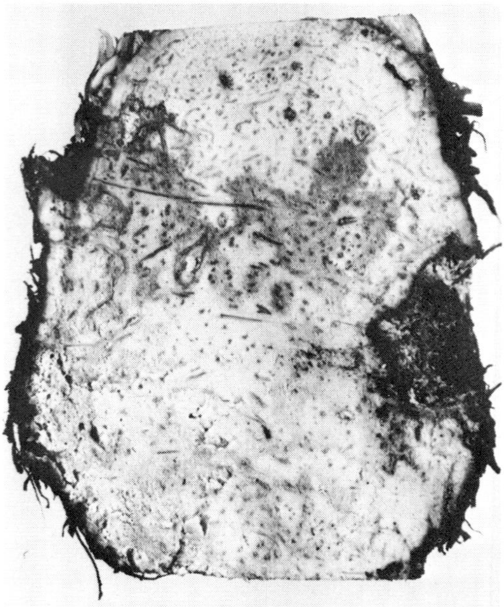

Fig. 140. Corm rot of the variety Samoa, after flooding.

Rotting is frequent in upland or dryland varieties immediately after floods. It can also occur during wet spells that follow a prolonged drought, or on land that has grown dalo for a very long time.

Symptoms
The leaves wilt. If the roots are dug, they appear completely rotted and the cut corms show a firm, watery, brown or purple rot extending usually from the base upwards (Fig. 140). In advanced stages the flesh crumbles like cheese and it has a very offensive odour.

Disease Cycle
Two species of the water mould *Phytophthora* are involved in corm rot. The swimming spores *(zoospores)* of the fungus swim about in soil water and attack roots usually after air has been cut off by flooding. Although dalo is generally considered a water-loving plant, some varieties are better suited to wetland conditions than others, and no variety can survive in stagnant water. Dryland varieties such as Samoa may develop corm rot if they are planted under wetland conditions.

Control

1. Grow the dalo variety best suited to the area: Kurokece, Sisiwa, and Vavai Dina are supposed to be grown under both dry and wetland conditions.
2. Plant dalo on well drained land, preferably on fertile slopes.
3. Rotate dalo if possible with other crops such as maize, rice, or bananas, that are not as susceptible to Phytophthora root rot.

4. If land has been flooded, transplant the dalo tops as soon as possible to better areas.

References: 56.

SHOTHOLE

Phyllosticta colocasiicola

Most varieties of dalo are susceptible to shothole but some, such as Sisiwa, show more than others. It is more evident in the wet zone. It does not appear to reduce yield.

Shothole is often confused with Phytophthora leaf blight which is common on wetland taro in Hawaii, but not known in Fiji.

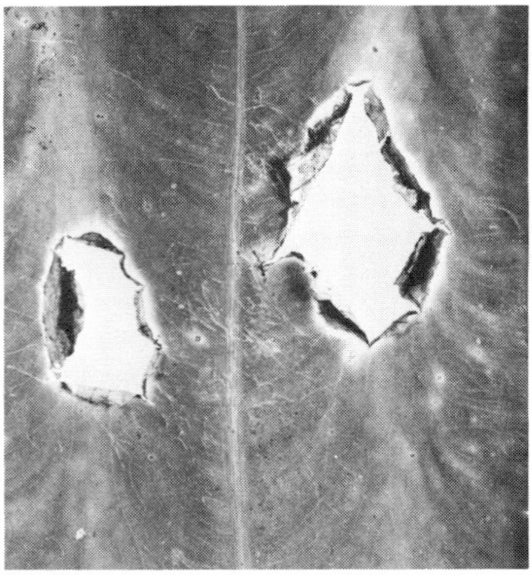

Fig. 141. Shothole on the Samoa variety.

Symptoms

Brown spots with yellowish-white haloes appear on the leaves. These spots enlarge to cover areas 1-2 inches in diameter, and the centres fall out (Fig. 141). Several holes may join to give the leaves a tattered appearance. Outer leaves are more badly damaged than the inner ones.

Disease Cycle

The disease is spread by spores produced in flask-shaped bodies *(pycnidia)* on the surface of the spots. The spores are pushed out during rainy periods and they splash on to the healthy inner leaves. Best conditions for infection occur during the cool season, particularly in shaded or very weedy locations. It is also probably more obvious during the cool season because dalo is growing slowly.

Control

1. Keep dalo plantations well weeded.
2. Do not plant wetland varieties under dryland conditions.

References: 56.

VEIN BANDING

Unknown

Fig. 142. Vein banding on the variety Samoa.

Fig.143. Severe interveinal banding on the variety Samoa.

A high percentage of the young plants of some varieties, such as Samoa, show this abnormality. Recently it was observed on a few plants of Daiga *(Amorphophallus campanulatus),* which is another edible plant related to dalo. Yield is apparently not reduced significantly.

Symptoms

Narrow white bands feather out from the main veins on the upper sides of the leaves (Fig. 142). Sometimes the interveinal tissues turn white. In rare severe cases, young plants are stunted and the leaves are somewhat tattered (Fig. 143). Older plants are relatively free of the condition.

Cause

The exact cause is unknown. It could be a hereditary and varietal condition, or it could be caused by a virus transmissible by either aphids or white flies, both of which are often found on dalo.

Control

1. Select planting material that is relatively free of symptoms.

FIJI DISEASE

Sugar cane Fiji disease Virus

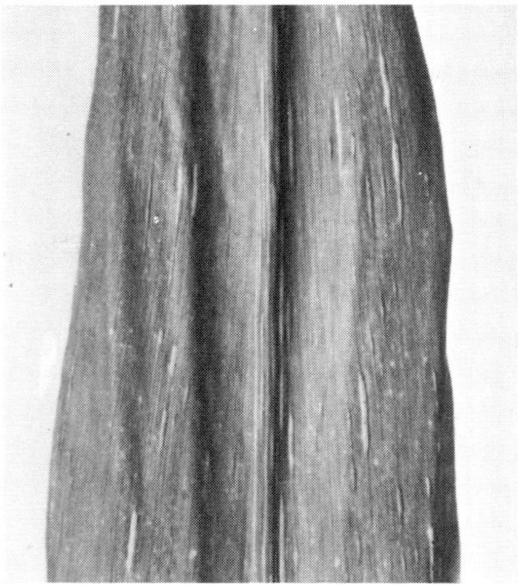

Fig. 144. Veinal galls on duruka leaf.

Fig. 145. *Left:* Stalk dwarfed by Fiji disease; *right:* healthy plant.

Fiji disease was first observed on sugar cane in Fiji in 1894. It was probably in duruka for many years before the sugar industry was established. It has been found on duruka everywhere it grows.

Symptoms

Symptoms on duruka are identical to those on sugar cane. Veinal galls (Fig. 144) varying greatly in length appear on the undersides of the leaves. When they are small they are the same colour as the leaf itself. Later they are greenish-white and easy to see.

Infected stools have a grass-like appearance and the leaves are dark green and very stiff and short (Fig. 145). During good growing conditions, infected stalks show leaf distortion, stunting, and the top looks like a fan.

Disease Cycle
The leafhopper *Perkinsiella vitiensis,* which is abundant on duruka, is known to transmit Fiji disease. It cannot be transmitted by the cane knife. The insect takes up the virus while it feeds on infected cane and then it injects it into the spindle leaves of healthy duruka. This host is extremely susceptible to the disease and it is an important source of infection for the crop.

Control
No special control measures are recommended because duruka is not a commerical crop at the present time. Duruka should not, of course, be grown in cane areas.

References: 55, 74.

LEAF SCALD

Xanthomonas albilineans

Leaf scald has been reported in Fiji but the occurrence of the disease is rare.

Fig. 146. Leaf scald (white streaks) in duruka.

Symptoms
White stripes (Fig. 146) with straight clearly defined parallel edges and a fine central red line extend the full length of the leaf blade. Side shoots and small suckers develop towards the bases of the stalks after the leaves wither and scald. Secondary shoots may also wither and die. If the stalks are cut, they show red streaks along the vascular bundles.

If the infection is severe, plants wilt suddenly and die, especially after dry weather. The wilting may be the only symptom.

Disease Cycle
The cause of leaf scald is a bacterium, *Xanthomonas albilineans,* that is transmitted by cuttings and cane knives.

Control
No special measures are recommended.

References: 55.

RED ROT

Glomerella tucumanensis

Red rot occurs on the midribs of duruka but it has not been found in the stalks. It is not serious.

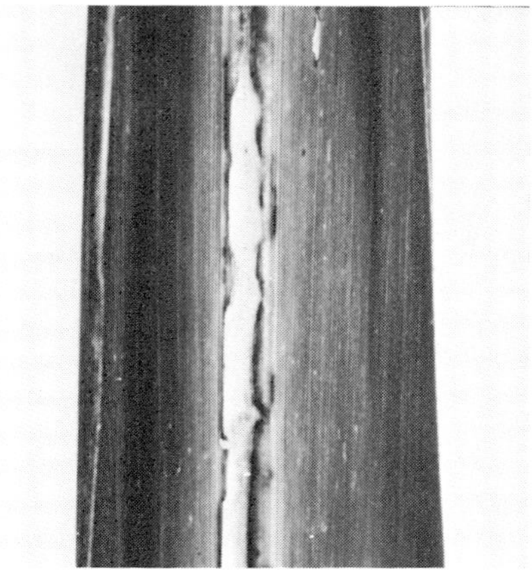

Fig. 147. Red rot on midrib of duruka leaf.

Symptoms

Small red spots appear on the midribs. They elongate to form long lesions the length of the leaf (Fig. 147), or they remain as distinct elongate spots. The colour of the spots change from bright red to straw-coloured in the centres, with red margins. Sometimes the spots are studded by black dots. In severe cases the leaf may split and wither.

Disease Cycle

The red rot fungus is very common on duruka and sugarcane. It is known in its conidial or asexual state as *Colletotrichum falcatum.* It is probably widespread in undecayed organic matter. It can get into the midribs of duruka through the feeding and egglaying punctures of leafhoppers. Other fungi may also invade these wounds. The fungus causing red rot is spread by *conidia* produced in great masses in the black dots *(acervuli)* during rainy periods.

The fungus also produces resistant spores *(chlamydospores)* that are thick-walled and permit it to survive during periods when mycelial growth is impossible. These spores can also spread the disease in cuttings.

Control

No special measures are recommended. Some duruka varieties may prove to be more resistant to red rot than others.

References: 55.

RING SPOT

Unknown

Ring spot is very common on duruka everywhere in Fiji. It does not appear to be serious.

Symptoms

Ring spot (Fig. 148) develops on the lower leaves. Spots are dark green to brownish, about ⅓-⅔ inch in diameter, and surrounded by a narrow yellowish zone. No death of leaves occurs.

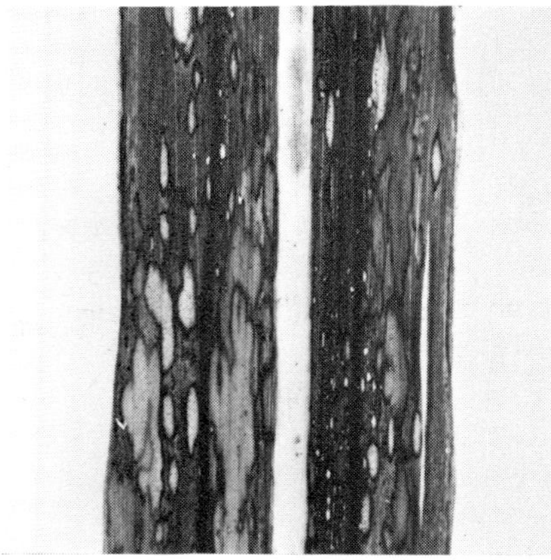

Fig. 148. Ring spot in advanced stage on duruka.

Disease Cycle

Although two fungi, *Leptosphaeria sacchari* and *Phyllosticta saccharicola,* have been found associated with ring spot, there is no proof that they can cause the spots independently. The evidence from work done elsewhere indicates that *Helminthosporium sacchari,* the pathogen of eye spot of sugar cane, is the primary cause, and that the other two fungi are secondary. The whole matter is an academic one, because the disease is quite unimportant.

Control

No measures are recommended.

References: 38.

RUST

Puccinia kuehnii

Rust is a very common but minor disease on duruka and on sugarcane.

Fig. 149. Rust on duruka.

Symptoms

Minute, elongated, yellowish spots develop on both leaf surfaces. They lengthen, turn brown to orange-brown, with a pale yellowish-green halo. In the last stages the characteristic reddish-brown rust pustules rupture the epidermis (Fig. 149). If there are many pustules, the leaves may die prematurely.

Disease Cycle

Wind and splashing rain spread the orange-coloured *uredospores.* The best temperatures for spore germination, and probably for disease development, are 20-25°F. *Teliospores,* which germinate to produce *basidiospores,* are known to be produced by *Puccinia kuehnii* only in Taiwan and Burma.

Control

No special measures are recommended.

References: 38.

YELLOW SPOT

Cercospora koepkei

Yellow spot is common on sugarcane and duruka. Some varieties of duruka are more susceptible than others.

Symptoms
Yellow spots about the size of pin-points (Fig. 150) appear on the young leaves. They later become irregular and run together. The spindle of duruka may be killed by the disease.

Fig. 150. Yellow spots on young leaf of duruka.

Disease Cycle
The fungus is spread by sickle-shaped spores *(conidia)* that are detached from the spots during wet, humid weather and high temperatures (28°C). The disease is very common on duruka and it may supply spores to infect sugar cane if it is being grown nearby. At least one duruka variety imported recently from New Guinea is quite susceptible.

Control
No special measures are currently recommended.

References: 55.

ANTHRACNOSE

Glomerella sp.

Anthracnose is not common. It is seen occasionally on yellow-fruited varieties. It occurs when fruit is left too long on the plants.

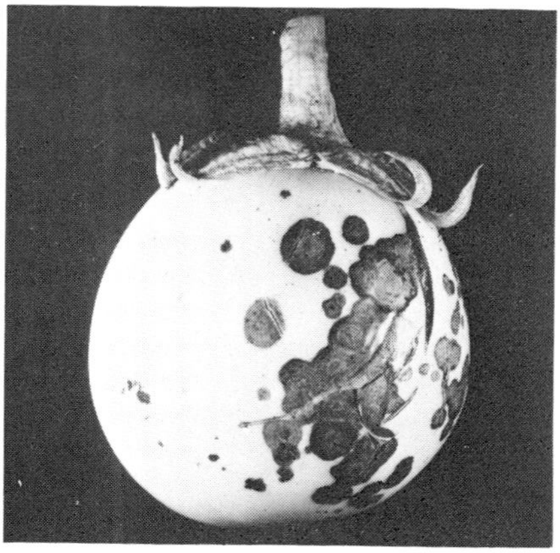

Fig. 151. Anthracnose on a yellow-fruited variety.

Symptoms

The disease starts as the fruit ripens, in the form of small black spots. They enlarge, sometimes in concentric rings, to spots $\frac{1}{4}$-$\frac{1}{2}$ inch in diameter (Fig. 151). Big spots may crack and expose the flesh. The spots are covered with black dots that exude a slimy pink mass during rainy weather. Fruit may drop to the ground, or it may develop a soft rot.

Disease Cycle

The fungus responsible for anthracnose probably attacks many other plants, including tomato and chilli. The disease is spread by the pink ooze of spores produced on the surface of fruit spots. The spores are splashed about by rain. The fungus is capable of producing spores also on crop refuse and it can live from one crop to the next on decaying fruit and stem parts. The disease is worst during wet warm weather and it develops best on overripe fruit.

Control

1. Harvest fruit at or slightly before maturity.
2. Rotate eggplant with non-solanaceous crops.

References: 14.

LEAF SPOT

Cercospora egenula, Corynespora cassiicola

Fig. 152. Cercospora leaf spot.

Fig. 153. Corynespora leaf spot.

Leaf spotting diseases are common but not important on eggplant. They seem to be worse during the cool season, but they never cause complete defoliation.

Symptoms
Leaf spots are light brown, circular, about $\frac{1}{8}$-$\frac{1}{4}$ inch in diameter, and sometimes they are bordered by a dark line (Fig. 152), or they are black and starshaped (Fig. 153). They usually occur on the lower leaves and become more numerous as the crop matures.

Disease Cycle
Leaf spot is caused by two fungi that are both spread by *conidia* produced on the surfaces of spots. They are washed off by rain and blown to healthy leaves. One of the fungi, *Corynespora cassiicola,* has a very wide host range. Some varieties of eggplant are more susceptible than others.

Control
No special measures are needed.

References: 14, 28, 86.

RUSSETTING

Unknown

A certain amount of russetting of fruit is seen on most soils in the Suva Peninsula of Fiji. It causes no apparent reduction in yield, but the internal browning associated with it may make the fruit unacceptable.

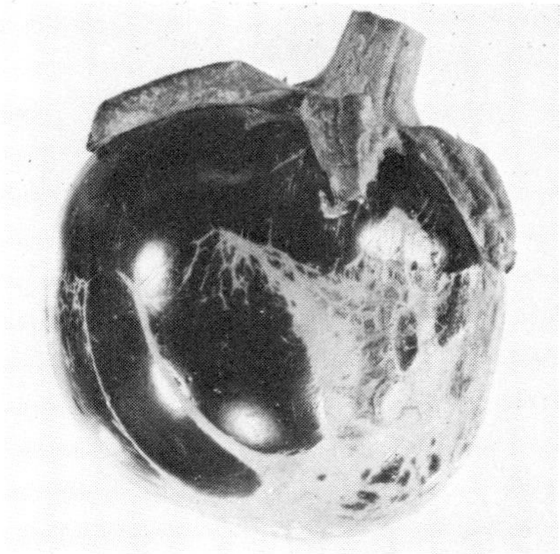

Fig. 154. Russetting of a purple variety of eggplant.

Symptoms
Fruits develop a rough skin at the blossom end (Fig. 154). This may happen on single fruits or on several on a single plant. Colour development on purple varieties is not normal, as the fruits remain green. If the fruit is cut, the flesh and seeds at the blossom end appear brown.

Cause
The cause is unknown. It is suspected that russetting is similar to blossom-end rot of tomatoes, because they both occur under the same environmental conditions, often in the same field. It is very common during drought or where the only fertilizer used is ammonium sulphate. It may be due to a moisture and nutrient imbalance.

Control
No measures can be recommended.

ROOT KNOT

Meloidogyne incognita

Fig. 155. Rhizome with lumps and small craters.

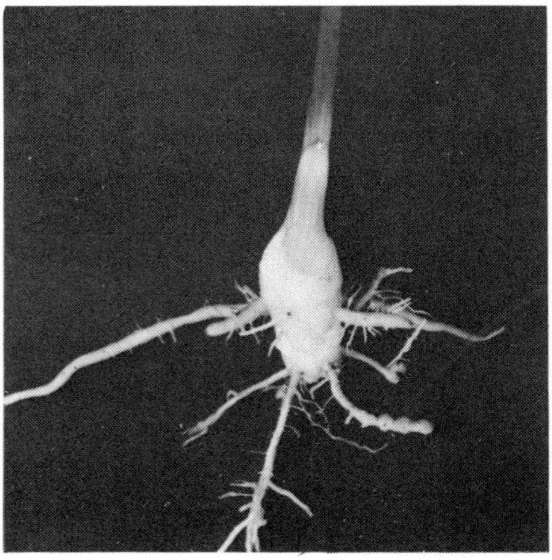

Fig. 156. Young ginger plant with bead-like swellings on the roots.

Root knot nematode infestations are responsible for most rejections of ginger for export. The yield may also be reduced.

Symptoms
The surface of the rhizome is lumpy and it shows small craters (Fig. 155). If the rhizome is cut, the flesh shows small, light brown, watersoaked areas about ¼ to ⅓ inch deep. Roots have characteristic swellings, like beads on a string (Fig. 156).

Disease Cycle
The root knot nematode species on ginger is the same as that on bananas, tobacco, tomatoes, carrots and many other hosts. The life cycle is also the same. Ginger grown for a number of years of the same land, or grown in rotation with the other nematode-susceptible crops, will be heavily infested.

Control

1. Rotate ginger with tolerant crops such as dalo, cabbage, maize, but *not* with tomatoes or carrots. Do not grow ginger for more than one crop on the same land.
2. Choose clean planting material.
3. Treat planting material by soaking in hot water at 122°F. for 10-15 minutes.
4. Fumigate land intended for ginger with Fumazone, by injecting an emulsion of 1 part of fumigant to 9 parts of water, at the rate of 5 c.c. per point, with points spaced 12 inches apart. Land must be in a very fine state of tilth before fumigation. Do not plant for at least 3 weeks.

References: 18, 19, 36, 79.

SOFT ROT

Pythium gracile, Erwinia sp.

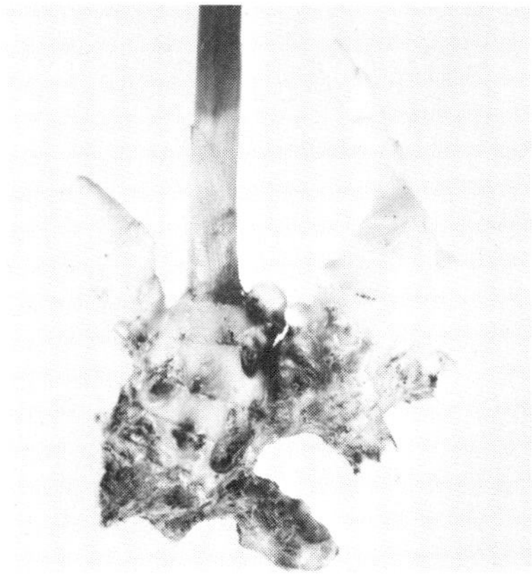

Fig. 157. Soft rot of young ginger rhizome.

Fig. 158. Mature rhizome which recovered from soft rot.

Soft rot is seen occasionally on ginger rhizomes intended for export. It seems to be more prevalent during the wet season.

Symptoms
Plants are yellowed and stunted. The rhizomes develop pockets of soft rot (Fig. 157). If the areas affected recover, the whole rhizome has scabby or depressed areas (Fig. 158).

Disease Cycle
Soft rot is initiated by the water mould *Pythium gracile.* Other species may be involved. The disease is favoured by a waterlogged soil, and by the use of diseased planting material. It is spread by *zoospores* that swim about in the soil water. Soft-rotting bacteria invading the rhizomes after the fungus also cause breakdown.

Control
1. Choose well-drained planting sites.
2. Avoid the use of infected rhizomes as planting material.

References: 36, 79.

ANGULAR LEAF SPOT

Phaeoisariopsis griseola

Fig.159. Round spots on a first leaf of haricot bean.

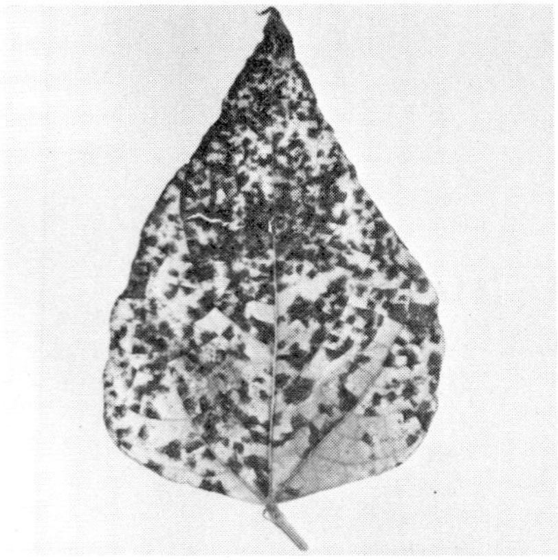

Fig. 160. Severe angular leaf spot on bean leaflet.

Angular leaf spot is very common during the cool season on haricot beans. It does not attack the pulses: mung, urd, cowpea, and long bean. It is not serious if fertility is maintained at a proper level, or if crop rotation is practiced.

Symptoms

The leaves are covered by angular brown spots (Fig. 160) about ⅛ inch to ¼ inch across. Petioles and pods may also be attacked. If the attack is very severe, beans may be defoliated. Spots on youngest leaves are round and sometimes ½ inch in diameter (Fig. 159).

Disease Cycle

Conidia produced on stalks above the leaf surface are splashed or blown from infected to healthy plants. The fungus may also be seedborne or it may live in bean refuse in the soil for as long as two years.

Best conditions for disease development are cool (68-75°F.) weather and frequent showers.

Control

1. Use a rotation in which beans do not follow each other for at least two years.
2. Spray with maneb (1 oz/3 gallons) or Kopi, Vitagran or Cuprox (2 oz/3 gallons).

References: 14.

HALO BLIGHT

Pseudomonas phaseolicola

Halo blight was first observed in Fiji in May, 1967. It has probably been here for much longer because bean seed is imported annually from New Zealand and Australia, where halo blight is common.

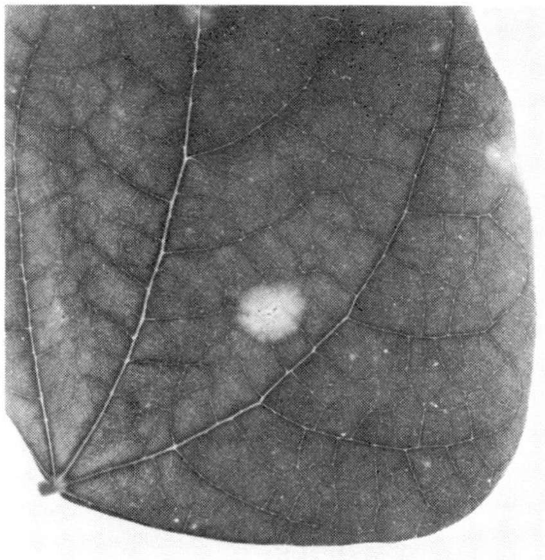

Fig. 161. Halo blight on dwarf bean leaflet.

Symptoms

Seedlings develop watersoaked spots on the stems just below the point of attachment of the cotyledons. Then they turn brown and die. Prominent yellow haloes (Fig. 161) develop around dark-green, shiny, angular spots on the older leaves. If the infection is severe, the whole leaf surface may turn yellow.

On stems, depressed, reddish-brown, elongated spots develop. An ooze is produced on their surfaces in wet weather. Pods show small, watersoaked circular areas that enlarge to become dark green and which also produce an ooze. Spots on mature pods are dark brown, but not sunken as with certain types of pod rot.

Disease Cycle

Halo blight bacteria are seedborne. The bacterial ooze is spread from infected seedlings by splashing rain and by insects. Leaves, stems and pods are infected when the bacteria enter stomata or wounds made by chewing insects such as grasshoppers, etc. If weather conditions are favourable, symptoms will show up within a week. The bacteria enter the seed by growing down into it through the vascular system of the pod, after getting in through the pedicel or through the pod wall. Then they are located in masses under the seed coat.

Bacteria may also be carried over from one bean crop to the next if crops are grown in close sequence.

Best conditions for spread of the disease are high temperature and high humidity.

Control

1. Insist upon a source of bean seed that can be certified disease-free.
2. Rotate beans with other non-legume crops.

3. Grow resistant varieties: *Dwarf beans:* Burnley selection of Canadian Wonder, Early White (Sydney Wonder), Feltham's Prolific, Pale Dun, Refugee; *Climbing beans:* Epicure, Kentucky Wonder, Market Wonder, White Kentucky Wonder.

References: 6, 14.

Bean Mosaic Virus, Cucumber Mosaic Virus

Beans are affected by a number of different kinds of mosaic which have not been distinguished from each other. Their effect on yield has not been accurately estimated.

Fig. 162. *Left:* mosaic-infected leaf; *right:* healthy leaf.

Symptoms

Leaves show a mosaic of light and dark green areas. The veins may be cleared or white in appearance and the leaf surface is usually severely wrinkled and reduced in size (Fig. 162). Certain types of mosaic make the plants bushy and stunted, and with a greater number of leaves than is normal. Sometimes the tops of plants are killed. The pods of some varieties, e.g., Bountiful, develop a greasy appearance.

Disease Cycle

It is suspected that common bean mosaic and cucumber mosaic viruses cause the disease in Fiji. Bean mosaic is seed- and insect-borne but cucumber mosaic is mainly insect-borne. Bean mosaic may survive for as long as 30 years in seed.

The best conditions for development of mosaic symptoms are temperatures of 70-80°F., and dry weather that promotes the build-up and movement of aphids.

Control

1. Ensure a source of seed that can be certified mosaic-free.
2. Grow resistant varieties, e.g., Seminole, Wade, Contender, Topcrop, Rival, Florigreen, Fullgreen, Idaho Refugee, Idaho Bountiful. As these have not been fully tested locally, some may prove to be susceptible to local types of mosaic.

References: 6, 11, 14.

POD ROT

Rhizoctonia sp.

Pod rot occurs everywhere beans are grown, under wet conditions at harvest time. It is not serious.

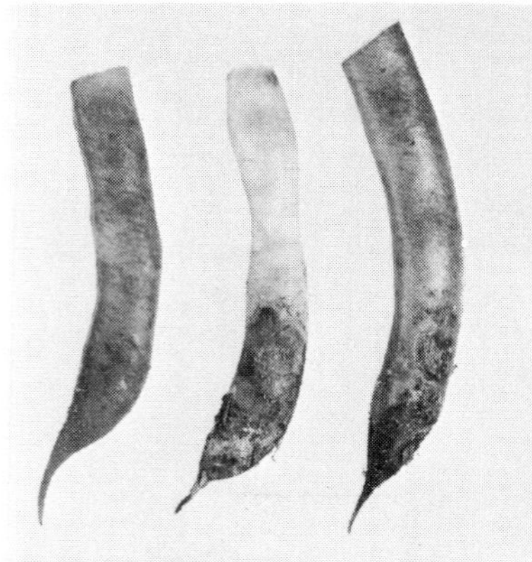

Fig. 163. Pod rot on haricot bean.

Symptoms
Cankers may develop at any point on the pods. They may be without any pattern (Fig. 163), or they may consist of concentric rings of grey and brown tissue. The area covered by the cankers may be ¾-2 inches in length, depending upon the size of the pods. Often black pinpoints are present on the surface of the cankers.

Disease Cycle
Pod rot is caused by a soilborne fungus called *Rhizoctonia,* which lives in undecayed organic matter. It can persist in soil for a very long time as *sclerotia,* or resistant bodies that look like black pinpoints. On certain wild legumes *(Mucuna platyphylla* L.*)* the fungus produces a target spot similar to the cankers on bean pods. The *sclerotia* or *hyphae* grow into the pods from soil particles splashed onto them during rain. Infection may also occur where pods are in direct contact with soil.

Control
1. Apply a leafy mulch to beans before they reach maturity.

References: 14, 62.

ROOT and STEM ROT

Fusarium sp. *Glomerella* sp.

A small amount of root and stem rot is seen in most bean plantations. It affects single plants here and there but it is not serious.

Fig. 164. Advanced root and stem rot.

Symptoms
There are gaps in the rows indicating that plants have died soon after emergence. Survivors may be dwarfed, yellowed, and without lower leaves. If the plants are dug, the tap roots appear dead (Fig. 164) and the lower stem may be hollow and rotted. In earlier stages, the tap root appears bright red. Lateral roots are killed back from the tips and they dry up entirely. Sometimes a pink or black fungus growth can be seen on the stems at soil level.

Disease Cycle
The fungi involved are both common inhabitants of undecayed organic matter and bean refuse. The spores are spread by running water, implements, animals, and seed. They attack seedlings usually through natural openings on the hypocotyls, or they get in through root tips and they grow up the root into the stem through the vascular system. When stems rot, the fungi produce spores on the rotted surfaces. The disease is worst where undecayed organic matter from a previous crop is left in contact with emerging seedlings. It is worst during the cool season when beans do not grow as rapidly.

Control
1. Treat seed before sowing with spergon (4 oz/100 lb or 1 tsp/1 lb).
2. Use a long rotation in which a cereal precedes beans.
3. Avoid susceptible varieties: Bountiful, Plentiful, Red Kidney, Refugee.

References: 6, 14.

Magnesium deficiency

Fig. 165. Interveinal yellowing on mature bean leaf.

Fig. 166. Complete yellowing and interveinal necrosis on older bean leaf.

Yellows, or interveinal yellowing, is common on haricot beans grown on acid soils low in fertility. It is often confused with insect damage and with certain virus diseases.

Symptoms

Mature and older leaves show an interveinal loss of chlorophyll (Fig. 165). The veins remain bordered by green. Lower leaves may be completely yellow, or even white, and the interveinal tissues turn brown and die (Fig. 166). Pod yields are reduced and plants die prematurely. They appear to be more susceptible to leaf-spotting diseases.

Yellow mosaic, with which yellows might be confused, causes an evenly-distributed mottling of the leaf surfaces. Also, yellows is fairly generally distributed in a plantation, and its occurrence can be correlated with similar symptoms on other crops grown in the same location.

Cause

The symptoms resemble those produced by a deficiency of magnesium on other legumes and tomatoes, and soil analyses have confirmed this suspicion. The soils upon which yellows was found are known to be fairly acid, a condition that makes magnesium unavailable.

Control

1. Check the soil fertility and use a fertilizer recommended by the Soil Chemist.

References: 14, 72, 84.

IMMATURE FRUIT ROT

Rhizopus artocarpi

This disease is common on jackfruit, but not on breadfruit. Elsewhere, in the Philippines, it is serious on breadfruit.

Symptoms
The young flowers and fruit develop a rot (Fig. 167) that advances very quickly. The fruit falls off the trees.

Fig. 167. Young jackfruit with fruit rot.

Disease Cycle
Observations on jackfruit in Fiji indicate that a species of borer attacks the young fruit and that a fungus, *Rhizopus artocarpi,* which is closely related to bread mould, colonizes the tunnels made by the insect. The fungus is a very common inhabitant of soil, and all kinds of decaying vegetation. Spread of the disease is by windborne spores.

Control
1. Bag the fruit before it becomes attacked by borers.
2. Remove and burn dead flowers and fruit.
3. Spray with Cuprox, Kopi, or Vitagran (2 oz/3 gallons) at 2-week intervals during the early fruiting season. Add DDT emulsion (1 oz/3 gal) to control the borers.

References: 56.

LEAF SPOT

Cercospora batatae

Leaf spot is common in several territories in the South Pacific. It is not known to reduce yields significantly.

Fig. 168. Leaf spot of kumala.

Symptoms

Angular, brown spots (Fig. 168) about 1/4 inch across develop on the leaves. If the attack is very severe, leaves may drop from the lower parts of the stems.

Disease Cycle

Leaf spot is caused by a species of *Cercospora* similar to that causing banana leaf spot. The life cycles of these fungi are similar, but unlike the banana fungus, the *Cercospora* on kumalas produces only *conidia.* These spores are spread about by rain and wind.

Best conditions for disease development are encountered during the cool season.

Control

No special measures are recommended.

References: 13, 14.

SCAB

Elsinoe batatas

This disease is fairly common in the South Pacific. It appears to be more severe during the cool season. It reduces growth severely in very susceptible varieties.

Fig. 169. The tip of a scab-infected kumala vine.

Symptoms

The terminal leaves are distorted, tattered, and reduced in size (Fig. 169). Scabby areas appear on the midribs, main veins, and petioles on the undersides of the leaves. These spots are grey to dark green in colour and they often crack crosswise. Similar lesions, sometimes over an inch in length, appear on the stems, which may also become twisted.

Disease Cycle

This has not been studied in detail. It is probably very similar to that of citrus scab, caused by a fungus which is closely related to *Elsinoe batatas.* Spores are produced on the surfaces of the scabby spots, from which they are washed and blown during rain to infect young, healthy kumala shoots. Best conditions for disease development appear to prevail during the cool season, with temperatures between 55°F. and 80°F.

Control

1. Experiment with maneb (1 oz/3 gal), applied every 10-14 days.
2. Grow resistant varieties.

References: 56, 70.

TUBER ROT

Geotrichum sp.   *Erwinia* sp.

Tuber rot of various types occurs wherever kumalas are grown but local losses have not been heavy.

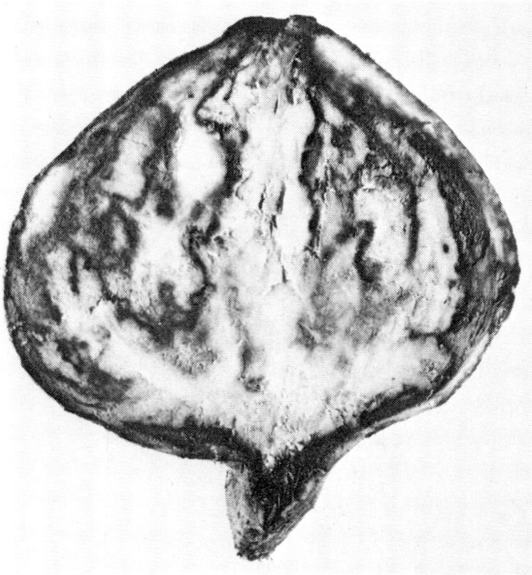

Fig. 170. Tuber rot of kumala.

Symptoms

Cut tubers show streaks and islands of blackened flesh (Fig. 170). They enclose grey and sometimes pink zones. The rotted areas give off a strong fermentative odour. Tunnels of insect larvae may be found associated with the rot.

Disease Cycle

Only storage-rotting fungi and bacteria have been isolated from samples of tuber rot collected recently from kumala fields. The insects found associated with the rot are also commonly found in storage. Perhaps field conditions are sometimes conducive to attacks by these organisms.

The possibility of involvement of species of *Pythium,* a water mould that thrives and attacks kumala roots and tubers under conditions of temporary flooding, cannot be excluded because they are difficult to isolate from advanced cases of rot.

Control

1. Plant kumalas on well-drained land.
2. Rotate with rice, maize, or sorghum.
3. Harvest tubers before they are overmature and store in a dry place.

References: 14.

DOWNY MILDEW

Sclerospora sacchari

Fig. 171. Downy mildew on maize leaf.

Fig. 172. Maize stunted by downy mildew.

Downy mildew is not very serious on maize but it is regarded as a threat to sugar cane where maize is a neighbouring crop.

Symptoms

Leaves show regularly spaced yellow and light green bands (Fig. 171) extending all the way from the bases to the tips. Leaves eventually turn brown and die out. Plants may be dwarfed (Fig. 172) and sucker more than is normal. Ears are poorly formed and multiple ears may occur. There is no shredding of the leaves. A white downy growth is sometimes visible on the diseased areas.

Disease Cycle

The downy mildew fungus on maize is the same as that on sugar cane. The disease is spread by spores formed on short stalks on the yellow and light green bands on affected leaves. These spores germinate by means of germ tubes, which penetrate the stomata. Resistant spores *(oospores)* may also be formed in leaf tissues, but they are not common and they are probably not important in spreading the disease. Spread is not very fast, because only single plants here and there appear to be infected.

Best conditions for mildew development are found in zones intermediate between wet and dry. Mildew is not found often in the wet zone. A certain amount of sunlight (day length) is required to mature the spores.

Control

1. Rogue and burn diseased plants.

References: 24, 55.

Nonspecific

Fasciation, or production of multiple, deformed ears occurs in maize every year. It is of very minor importance.

Fig. 173. Fasciation in maize.

Symptoms
Several ears are produced at the same location on a plant (Fig. 173). They are twisted and reduced in size, and they emerge from about the same point. Production of grain is severely limited or absent.

Cause
The condition may be associated with a number of causes, all of which injure and disturb the growing point of the ear during the embryo stage. These causes are suspected to be the accidental application of herbicide (MCPA or Dalapon) during an early stage of growth, lack of pollination or infection by downy mildew or mosaic diseases. Aphid feeding could also be a cause. A certain amount of fasciation may be inherent in certain maize varieties. There is no evidence that the condition spreads from plant to plant.

Control
No special control measures are recommended.

INTERVEINAL CHLOROSIS

Magnesium deficiency

This disorder is very common in maize but it is doubtful that it really reduces yield. It may have some effect on the food value of maize meal.

Symptoms
During all stages of growth maize plants show white or light yellow bands between the veins (Fig. 174).

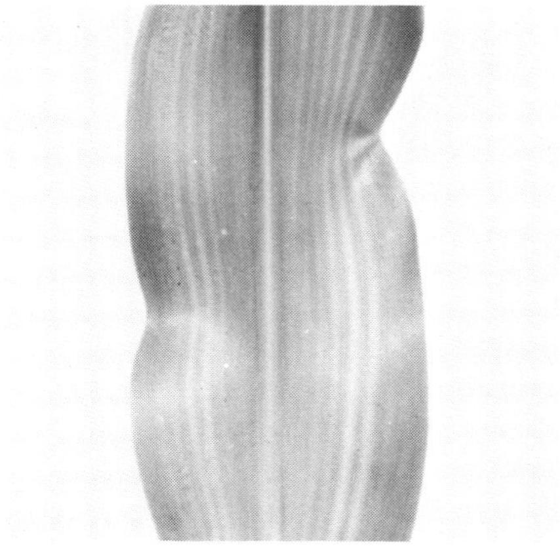

Fig. 174. Maize leaf with interveinal chlorosis.

Cause
The cause is generally attributed elsewhere to a deficiency of magnesium. It seems to be common on very acid soils, such as those in the lower Rewa Valley of Fiji.

Control
1. Apply magnesium sulphate or dolomitic limestone, according to the directions of the Soil Chemist.

References: 84.

LEAF SPOT

Cochliobolus heterostrophus

Leaf spot is a minor disease of both seedlings and mature plants. It is difficult to estimate its importance because rust is so very much worse.

Fig. 175. Leaf spot on mature maize leaf.

Symptoms

Spots on seedlings are round to elongate, about $^1/_8$-$^1/_4$ inch long and on mature plants they are elongate, about the same distance across ($^1/_8$ inch) and $^1/_2$ inch long (Fig. 175). On seedlings the spots have grey centres and purple edges and on mature plants they are grey with indistinct edges.

Disease Cycle

The primary source of infection is spores from crop refuse. Spores produced on the leaf spots spread the disease within the crop.

Seedlings may become susceptible to leaf spot as a result of imbalance between nitrogen and potassium. Best conditions for disease development occur during cool, wet weather.

Control

1. Check soil fertility.
2. Plough under maize and rice refuse immediately after harvest.
3. Grow a resistant variety. Tuxpeño appears to be resistant.

References: 24, 39, 40, 60.

MOSAIC

Cucumber Mosaic Virus, Sugar Cane Mosaic Virus

Mosaic is a common but relatively unimportant disease. It affects only single plants here and there and it does not appear to spread in the maize crop.

Symptoms

The leaves are striped by broken dark green, light green, and yellow lines (Fig. 176). Plants may be stunted. Formation of ears is not normal and the yield from infected plants may be low or nil.

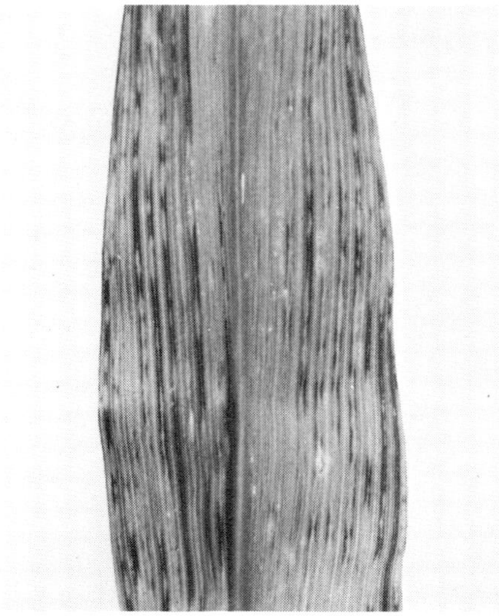

Fig. 176. Mosaic on the variety Local Yellow of Fiji.

Disease Cycle

Two viruses, cucumber mosaic, transmitted by the melon or cotton aphid *(Aphis gossypii)*, and sugar cane mosaic, transmitted by the maize aphid *(Aphis maidis)*, may be involved in maize mosaic. No experimental work has been done locally to sort out the complex. The source of virus for maize is probably duruka, wild grasses, and certain weeds and vegetables upon which these aphids feed.

If seedlings are infected soon after emergence, they remain dwarfed and are probably self-eliminating. If mature plants are infected, they may show little or no abnormal reaction. High percentages of infected plants can be correlated with high aphid populations.

Control

1. Keep weeds under control.
2. Spray seedlings with malathion (1 oz/3 gallons).
3. Rogue and burn diseased plants.

References: 24, 55, 74.

PURPLING

Phosphorus deficiency

Fig. 177. Maize seedling showing purpling and a poor root system.

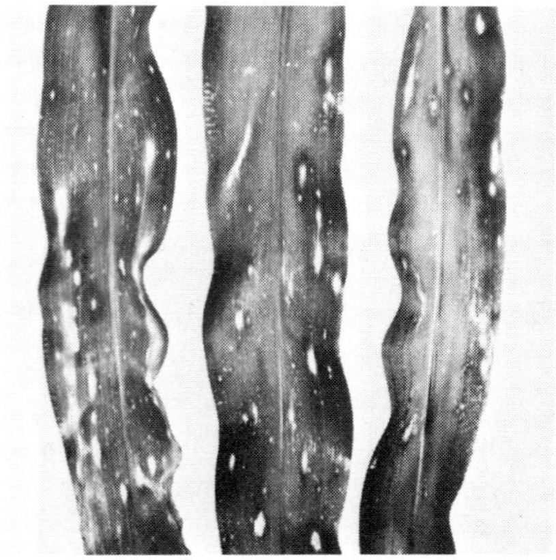

Fig. 178. Seedling maize leaves showing purpling and leaf spot.

Seedling maize often shows this disorder, particularly if sown in dry soil or soils that are low in fertility.

Symptoms
The stalks and leaves of seedlings show a pronounced purple colour (Fig. 177) in varieties that are normally green. When maize attains the height of a foot or more, the purpling tends to be replaced by normal green. Root systems are often under-developed (Fig. 178) and they may become rotted.

Cause
During the early stages of growth, the maize plant is quite demanding of phosphorus, which is slowly available in many soils in the South Pacific. If a drought occurs at sowing time, the level of available phosphorus may be deficient. It is also rendered unavailable by a high soil acidity.

Control
1. Apply a side dressing of superphosphate and possibly lime at sowing time, according to the recommendations of the Soil Chemist.

References: 39, 84.

RUST

Puccinia polysora

The most serious disease of maize and sweet corn in the South Pacific is rust. Fortunately the Tuxpeño variety of maize that was introduced into Fiji is highly resistant. However, there is no assurance that other rust races will not appear eventually to attack this variety.

Fig. 179. Rust on Fiji or yellow maize.

Symptoms

The disease begins as translucent or watersoaked pinpoints. These develop into round or elongated reddish-brown pustules on both surfaces of the leaves. The pustules on ears and stalks are about ¼ inch long and they occur in clusters (Fig. 179). Ears from affected plants are small and they may not mature properly.

Disease Cycle

Uredospores produced in the pustules *(uredia)* spread the diease. These spores are very light and they are spread by wind and rain. When they land on healthy maize leaves, they produce germ tubes that penetrate the leaves through the stomata. On resistant varieties such as Tuxpeño, pustule development is arrested, and the symptoms do not get far beyond the fleck stage.

Rust is worst in a moist warm climate.

Control

1. Use resistant varieties such as Tuxpeño or others that are available from the Rockefeller Foundation, Mexico.
2. Plant maize during the cool dry season.

References: 24, 36.

SHEATH and PINK EAR ROT

Gibberella fujikuroi, Gibberella zeae

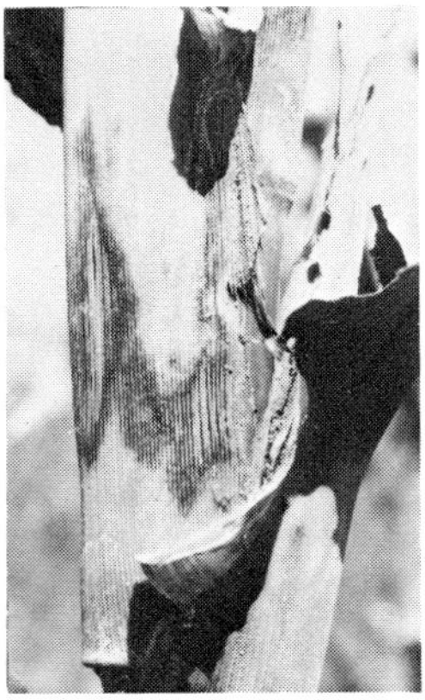

Fig. 180. Sheath rot.

Fig. 181. Pink rot on Tuxpeño.

Incidence of this disease varies from year to year. It is worst during wet warm weather and it makes maize unsuitable for feeding to livestock.

Symptoms

Purplish-black, red, or pink rot commences at the bases of the leaf sheaths, or at the nodes (Fig. 180). They are sometimes covered by a white fungus growth. Stalks may break if badly affected.

A pink or reddish-brown rot also develops at the tips of the ears (Fig. 181) and progresses downward. Kernels may germinate prematurely on the ears, or they may germinate poorly and develop seedling blight.

Disease Cycle

Two species of fungi are associated with sheath and ear rot. They are soilborne and they are not limited to maize. They grow on cereal crop refuse and grasses. They are spread by sickleshaped *conidia* of the Fusarium type or more rarely by *ascospores* produced in blue-black *perithecia* on the rotted surfaces of leaf sheaths. The fungi grow down the dried tassels or silks into the kernels.

If the soil is cold and wet, seed rot and seedling blight is common from infected kernels. Ear rot is worst when wet warm weather occurs at harvest. The disease may spread from kernel to kernel if the grain is stored under humid conditions.

Maize varieties with tight husks and with ears that turn down at maturity are somewhat tolerant to pink rot. There is no real resistance.

Meal with a high percentage of rotted maize kernels has been known to cause occasional abortion in pigs and stomach upsets in cattle.

Control

1. Plough maize stubble and refuse down to a depth of 9-10 inches immediately after harvest.

2. Turn the ears down as soon as they are mature to allow them to shed the rain.
3. Rotate maize with non-cereal crops.
4. Store maize under dry conditions.

References: 24.

WHITE LEAF

Unknown

This disorder was observed first in the Rewa Valley of Fiji during 1964, and it has recurred. The plants suffer a premature loss of leaves.

Symptoms

When the plant approaches maturity, the edges of the leaves turn white and shred (Fig. 182). The youngest leaves may be deformed and completely white. The condition can be confused with mosaic, but the virus disease is confined to single plants. White leaf may affect all the plants in a field.

Fig. 182. White leaf of maize.

Leafhopper injury also resembles white leaf but it can be distinguished by the short longitudinal slits marking the sites of egglaying in the midribs and along the veins.

Cause

Elsewhere a similar disorder has been attributed to a zinc deficiency that is common in very acid soils. Similar symptoms may also be caused by deficiency of potash. The loss of normal green colour is caused by an interference with synthesis of chlorophyll.

Control

1. Check the pH of the soil and correct according to the recommendations of the Soil Chemist.

References: 84.

ANTHRACNOSE

Glomerella cingulata

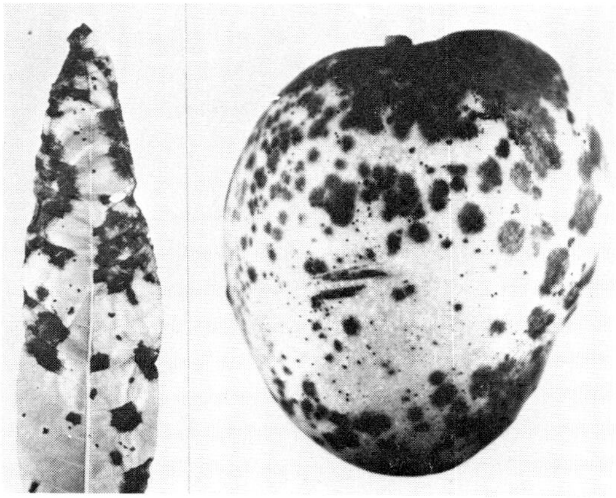

Fig. 183. *Left:* Anthracnose spots on mango leaf; *right:* spots on ripening fruit.

Anthracnose is the most serious disease of mangoes in the South Pacific, particularly in zones with a high rainfall.

Symptoms
Blossoms, young fruit, twigs, and small branches are killed. Dark brown, circular or angular spots that later turn purple develop on the fruit (Fig. 183). Spots on the leaves are angular at first and about ⅛ inch across. In severe cases they are irregular, like those on the fruit (Fig. 183).

Disease Cycle
Pink masses of spores are produced on the spots on the leaves, twigs, and fruit during very wet weather. They are spread by splashing rain and wind. If the weather during flowering is wet, the fruit set is very low or it does not occur at all, because the blossoms become infected. Fruit infection occurs when it is very green, and spots develop when the fruit starts to ripen.

Control
1. Grow mangoes commercially in areas with an annual rainfall of 60-70 inches.
2. Starting at flowering time, spray the trees with maneb (½ lb/2 gal) by misting machine every 7-10 days for a month. Add 1 tbsp of X-77 or Agral LN to the water before mixing in the fungicide.
3. Immerse fruit in hot water for 20 minutes at 118°F. before shipping.

References: 36.

NECK ROT

Glomerella spp., *Aspergillus flavus, Aspergillus niger*

Under South Pacific conditions smudge or neck rot prevents onions from maturing and curing properly. It can also cause serious losses in storage.

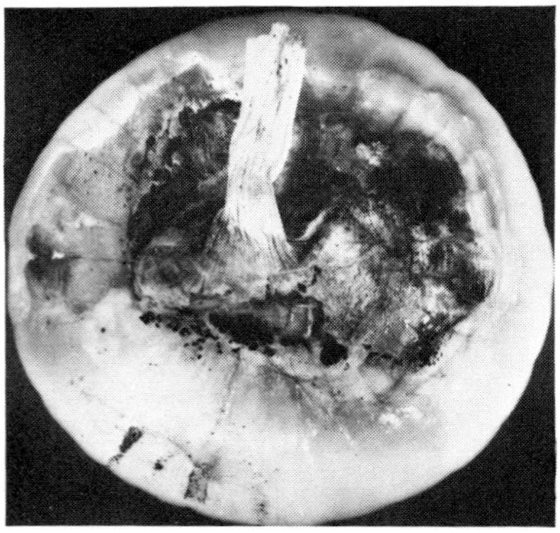

Fig. 184. Neck rot on a white-skinned onion variety.

Symptoms

Dark green and black spots appear on the necks of the bulbs (Fig. 184). They are usually followed by a soft rot.

Disease Cycle

As the crop approaches maturity, the outer layers of skin on the necks die and they are attacked by weakly parasitic species of *Glomerella.* In Fiji two species of *Aspergillus* may also be involved. These fungi are very common on all sorts of decaying organic matter. They are spread by *conidia* blown about by wind. After the fungi have invaded the neck tissues, soft-rotting bacteria move in and invade the bulbs. The result is a stinking mess.

Wet warm weather at harvest favours development of neck rot. Thrip damage to the neck tissues may also contribute. White skinned onions are more susceptible than the yellow and purple varieties.

Control

1. Grow onions in the drier zones.
2. When onions have reached a marketable size, bend the necks over.
3. Harvest during dry weather.
4. Store onions by hanging them in a dry place or by placing them one layer thick in slatted crates.
5. Experiment with yellow and purple varieties.

References: 14.

Virus

Rosette is definitely the major factor limiting the successful introduction of pangola grass in Fiji. It has decimated pastures wherever they have been established.

Fig. 185. *Left:* Healthy pangola plant; *right:* plant with rosette symptoms.

Symptoms

About 2 years after a pangola plantation has been established, individual plants become stunted and purple. They are easily uprooted by grazing animals. If fertilizer is applied to the paddocks, a temporary recovery occurs. Under glasshouse conditions, infected plants show interveinal yellowing, twisting, stunting, and multiple tillering (Fig. 185). Veinal galls similar to those of Fiji disease of sugarcane have been observed.

The plantation is finally invaded by more aggressive grasses and pangola dies out.

Disease Cycle

Rosette is very similar to pangola stunt, or rosette, described from Surinam, Florida, and Taiwan. It has been shown in Surinam that a leafhopper transmits the disease, suspected to be caused by a virus. It is not known whether rosette was brought into Fiji or existed here in local grasses previous to the introduction of pangola.

Control

1. Replace pangola by a more suitable grass.
2. Experiment with the resistant A-24 strain of pangola, if it is available from a reliable source.

References: 25, 51.

CORKY SPOT

Unknown

Complaints are occasionally received about corky spot. It is apparently prevalent at certain times of the year.

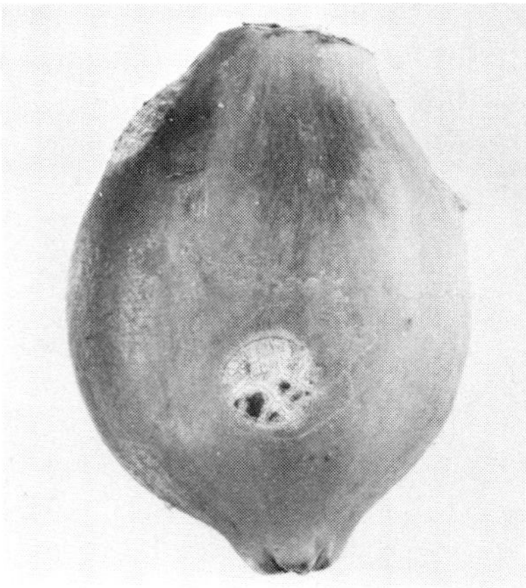

Fig. 186. Corky spot on young papaw fruit.

Symptoms

Ripe papaws show corky areas about ½-1 inch in diameter in the flesh (Fig. 186). They match sunken areas of equal size on the outside of the fruit. Young green fruit may show raised corky areas that later become sunken.

Cause

The cause is unknown but sucking bugs, which are known to feed on immature fruit, are under suspicion.

Control

No special measures are recommended.

FRUIT and ROOT ROT

Phytophthora nicotianae var. *parasitica*

Fig. 187. Collar and stem rot of papaw.

Fig. 188. Fruit rot on papaw.

Fruit and root rot is quite common in wet areas on poorly drained land. It can ruin a crop completely.

Symptoms

A firm rot develops on the green fruit, starting usually from the stem end. The fruit becomes covered by a white, fluffy mycelial growth (Fig. 188). It then shrivels and turns black. Water-soaked cankers may also develop on the stems.

Roots develop brown lesions. The rot extends from there into the base of the stem at ground level (Fig. 187). When the root are completely rotted the plants topple over. Seedling papaws may damp off or develop a jelly-like basal rot.

The leaves of infected plants turn yellow and drop leaving a very small crown.

Disease Cycle

The species of *Phytophthora* associated with papaw fruit and root rot may be the same as that causing pineapple heart and root rot, dalo corm rot, and passion fruit collar rot. The conditions favouring development of these 4 diseases are basically the same. (See dalo corm rot).

Control

1. Plant papaws in well-drained land. Plant on mounds in wet areas.
2. Avoid land previously planted to papaws, pineapples, passion fruit, particularly if these crops have had the same kind of trouble.

3. Fumigate seedbed soil with methyl bromide or plant seed directly in the field.
4. Remove and burn diseased trees, fruit, and roots.
5. Apply maneb (1 oz/3 gallons) or Kopi, Cuprox, or Vitagran (2 oz/3 gallons) weekly to the fruit during very wet weather.

References: 21, 35.

Botryodiplodia theobromae, Fusarium, Glomerella spp, *Rhizopus* sp.

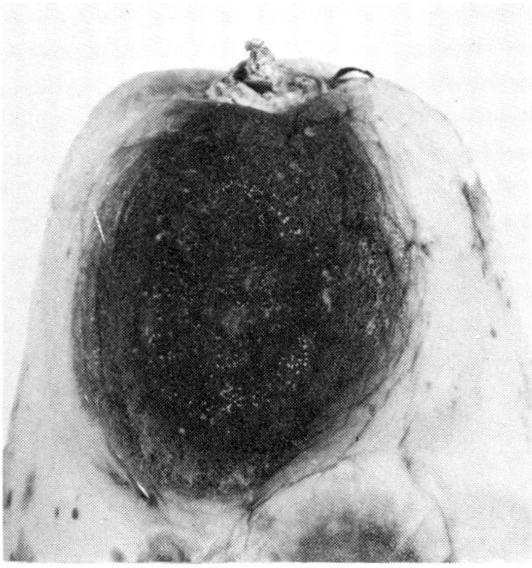

Fig. 189. Ripe fruit rot caused by *Botryodiplodia theobromae.*

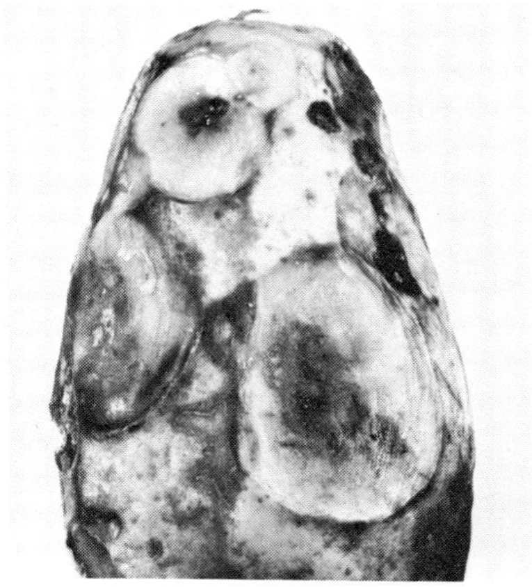

Fig. 190. Rot caused by a species of *Glomerella.*

Ripe fruit rot is very common everywhere and it is responsible for market deterioration of fruit. If papaws became an export crop it will have to be controlled.

Symptoms
Round, sunken, watery spots varying in size from ½-1½ inches in diameter appear anywhere in the ripening fruit (Figs. 189, 190). The spots are covered with masses of pink, black, or grey spores. Eventually the entire fruit collapses.

Disease Cycle
The fungi responsible for ripe fruit rot are weak parasites that abound in soil, dust, and plant refuse. They infect the fruit often before ripening and remain latent.

Infection may occur through wounds or lenticels. Best conditions for disease development are encountered during exposure to wind, driving rain, and rough handling.

Control
1. Pick the fruit green and allow it to ripen off the tree.
2. Dip fruit after harvest in hot water at 116°F. for 20 minutes.

References: 35, 66.

SHOTHOLE

Corynespora cassicola

Shothole is a very common leaf-spotting disease in the South Pacific but it is not important.

Symptoms
Light brown circular spots about ¼-⅓ inch in diameter appear on the lower leaves first and gradually spread upwards. The centres of the spots often fall out to produce a shothole effect (Fig. 191). Spots on petioles are oval in shape and they are covered by a dark brown growth of fungus. Spots on the fruit are small, dark brown, and sunken. They do not appear to cause any rotting.

Fig. 191. Papaw leaf with shothole disease.

Disease Cycle
The shothole fungus has a very wide host range. In Fiji it has been found also on tomatoes and eggplant. It is probably soilborne for part of its life cycle. Spores are produced on long slender stalks on the surface of leaf spots and they are spread by wind and rain. Infections on young leaves and fruit are probably arrested until the tissues start to age. The disease may be worse on poor soils.

Control
No special control measures are recommended.

References: 28, 86.

ANTHRACNOSE

Glomerella phomoides

Fig. 192. Anthracnose with brown spot, on purple passion fruit.

Fig. 193. Anthracnose on overripe yellow passion fruit.

Anthracnose is found in association with brown spot and with mosaic (woodiness). It is not in itself an important disease but it indicates a weakened condition.

Symptoms

The leaves show ashy white spots about ⅛-¼ inch in diameter. The fruit may also be invaded. The spots are raised on purple passion fruit (Fig. 192) and they may cover most of the fruit. They are covered by black pinpoints (Fig. 193) and under wet conditions a pink mass of spores is evident. Overripe fruit often develops severe anthracnose after it falls off the vines (Fig. 193).

Disease Cycle

The anthracnose fungus is very common on ripening fruits of all kinds, and also on woody tissues in an early stage of decay. Virus infection may kill out spots of leaf tissue, which are then invaded by *Glomerella.* The disease is spread by *conidia* of the Colletotrichum state. These spores are produced in slimy masses in the black pinpoint-like *acervuli* on the surfaces of spots. Splashing rain spreads them about.

Plants weakened by low fertility, collar rot, virus infection, etc., are more susceptible to anthracnose.

Control

1. Keep up soil fertility.
2. Keep other diseases and insect pests under control.

References: 65.

BROWN SPOT

Alternaria passiflorae

Brown spot is common in the Sigatoka Valley of Fiji but it is not yet of sufficient importance to justify a spray program. It affects only the rind of the fruit and not the pulp.

Fig. 194. Brown spot on yellow passion fruit.

Symptoms

Light brown circular sunken areas appear on the sides of the fruit (Fig. 194). They enlarge and they may shrink the fruit. They are often covered by reddish-brown spores. Spots on the leaves are $\frac{1}{4}$-$\frac{1}{2}$ inch in diameter, at first light brown and then drying out in the centres.

Disease Cycle

Reddish-brown spores are produced on the spots on the fruit. They may also be produced on old fallen leaves. These are spread by rain and wind to healthy fruit and leaves. It is quite possible that some infection of passion fruit occurs from other plants and from undecayed organic matter in and around plantations. The best conditions for infection are found in wet warm weather. Unpruned, over-crowded vines are most susceptible to severe attacks.

Control

1. Train vines properly and prune at the right time to avoid overcrowding.
2. Take ripe fruit to the processor as soon as possible. Do not leave it on the ground.
3. Spray in severe cases with maneb (1 oz/3 gallons) every 10-14 days.
4. Avoid a fertilizer imbalance between N and K.

References: 65.

Phytophthora nicotianae var. *parasitica, Phytophthora cinnamomi*

Fig. 195. Bark splitting at the base of the vine.

Fig. 196. Bark peeled away to show rotting.

Collar rot is found in all plantations over a year old. It is serious during the wet season and also just after floods. It is the main reason for replanting vines.

Symptoms

Vines wilt and turn yellow and drop their leaves. The bark at the bases of the vines just above the soil line splits (Fig. 195). If the bark is peeled away, a reddish-brown rot (Fig. 196) can be seen progressing upwards.

Disease Cycle

Two species of *Phytophthora* or water mould are involved in collar rot. They are worst in poorly drained land because they thrive apparently without oxygen.

The fine roots are invaded by the swimming spores *(zoospores)* of the fungi. Infection then progresses upwards into the collar of the vine along the inner bark layers. Some infection may also occur through injuries caused by cane knife cuts at the soil line. The fungi involved can attack many plants and they are always present in soil.

The purple variety of passion fruit appears to be more susceptible than the yellow one under local conditions.

Control

1. Plant vines on mounds and avoid cane knife cuts during ringweeding. Try chemical weeding close to the vines.
2. Replant a new area at the end of the first or second year, depending upon losses.
3. Rotate passion fruit with maize, sorghum, peanuts, rice.

References: 65.

FRUIT CRACKING

Disturbed water relations

Fruit cracking is common in the Sigatoka Valley and elsewhere in Fiji, usually at the beginning of the wet season.

Symptoms
Fruit of any stage of growth develops long cracks running down and across (Fig. 197). If the fruit is very young, the cracks become severely scabbed, and it may not develop further. Mature fruit may be used for juice because the cracking is limited to the rind.

Fig. 197. Cracking of yellow passion fruit.

Cause
Cracking is attributed to sudden changes in water supply at times when the fruit is still growing rapidly. Usually it occurs at the end of the dry season, but any period of drought followed by abundant rain can also induce it.

Control
No special measures are recommended as it is not a practice now to irrigate passion fruit.

Cucumber Mosaic Virus Woodiness Virus

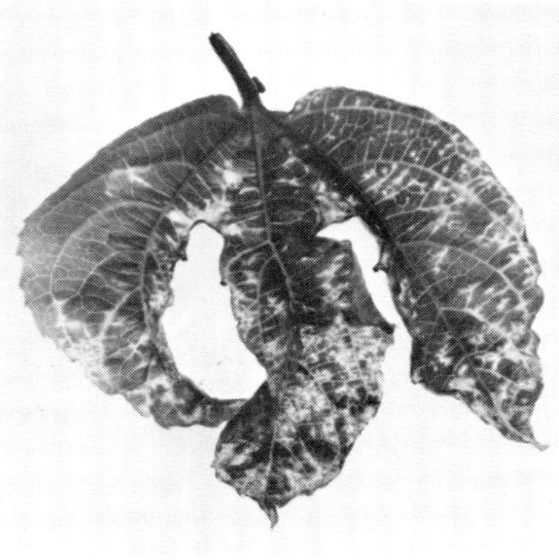

Fig. 198. Mosaic on yellow passion fruit leaf (above left).

Fig. 199. Ringspot pattern on yellow passion fruit (above right).

Fig. 200. Woodiness of purple passion fruit (left).

Mosaic and woodiness were first detected on passion fruit in the Sigatoka Valley of Fiji during 1965. It is probably widespread in the South Pacific, because the cucumber mosaic virus has many hosts. The disease appears to be worst just after the cool season. Losses have not been serious.

Symptoms

The young leaves may show a faint or pronounced light green or yellow mottling, or a mosaic pattern (Fig. 198). Sometimes shot-holes and ring spot pattern (Fig. 199) develop. Leaves may also be deformed and undersized. At certain times of the year, and on poor soils, fruits show symptoms of woodiness. They are undersized, the rind is woody and hard to cut, and the surface may be roughened and covered with corky outgrowths (Fig. 200). The pulp is reduced or entirely absent.

Disease Cycle

Two viruses, the cucumber mosaic virus, and the passion fruit woodiness virus, may be associated with this disease in Fiji and elsewhere. The cucumber mosaic virus is known to have many hosts among other crops and weeds. The viruses are transmitted by the handling of plants, by pruning shears, and by aphids. Infection of passion fruit vines is inevitable.

Best conditions for expression of woodiness are encountered during the winter months, but symptoms do not usually show until October-December when the vines are flowering and fruiting heavily. Warm weather masks symptoms.

Control

1. Keep weeds under control in and near plantations.
2. Grow passion fruit seedlings as far as possible from cultivated passion fruit and cucurbits.
3. Replant vines every 2 years to compensate for decline due to virus infection. Replanting will have to be done anyway to replace losses due to collar rot.

References: 23, 41, 54, 65, 74.

THREAD BLIGHT

Corticium solani

Thread blight occurred on passion fruit in the Sigatoka Valley for the first time in 1966, when unpruned vines were affected.

Symptoms
Leaves show big brown or black patches. They may drop or they may remain suspended from the vines. Fine brown threads can be seen running along the vines and down on to the leaves. A fine pink fuzz of spores may also be seen on the leaves. Severely affected leaves appear shredded (Fig. 201).

Fig. 201. Leaf severely affected by thread blight.

Disease Cycle
The thread blight fungi are very common in the tropics on many hosts. They attack rubber, tea, coffee, cocoa, and other plantation crops. They are probably normal inhabitants of the dead outer bark of trees but they can invade living tissue. Conditions favouring the disease are excessive shade, heavy rainfall, and heavy vine growth.

The fungus kills the leaf tissues immediately underneath its growth. Spores *(basidiospores)* produced on stalks *(basidia)* arranged in a thin layer on the superficial weblike growth may spread the disease from one area to another.

Control
1. Prune the vines of affected parts and of excessive growth.
2. Spray a couple of times with Kopi, Cuprox, or Vitagran (2 oz/3 gallons) to kill the basidiospores.

References: 7.

BACTERIAL WILT

Pseudomonas solanacearum

A few plants affected by bacterial wilt were found at Dobuilevu, Viti Levu, Fiji, in 1965. The Local Spanish variety was worst affected. The disease has been known for years in Indonesia. In Fiji it is not serious on peanuts, but it makes the crop a risky one in rotations including tomatoes, potatoes, and tobacco.

Fig. 202. Wilted plant of Local Spanish variety.

Symptoms

Plants turn yellow and collapse (Fig. 202). If the stem is cut across at soil level, a grey slimy ooze can be pressed out of the vascular ring.

Disease Cycle

The pathogen of bacterial wilt in peanuts is the same as that affecting solanaceous crops and weeds. It has probably been in the South Pacific for a long time before the introduction of peanuts. The other legume occasionally affected is cowpea.

Control

No special measures are recommended other than to plan rotations that avoid solanaceous crops immediately before peanuts.

References: 24, 42.

CROWN ROT

Aspergillus niger

Crown rot has been observed in the Sigatoka Valley, where it causes numerous gaps in the rows. It is associated with seed rot. The two diseases may reduce stands by as much as 25% depending upon the variety. Mani Pintar is more susceptible than Virginia Bunch or Local Spanish.

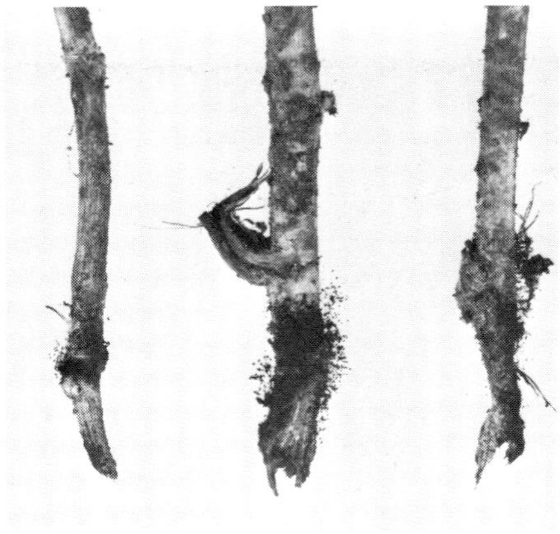

Fig. 203. Rotted stems with a mouldy growth.

Symptoms

Plants wilt before the pegging stage and they can be pulled easily. The crown shows a rot extending from the point of attachment of the cotyledons up to and sometimes above the soil line. Often a black mould can be seen on the rotted area (Fig. 203).

Disease Cycle

The crown rot fungus is both soil and seedborne. It attacks injured pods and it becomes established on the seed itself. When it germinates, the fungus may kill the young seedling before it emerges from the soil, or it may attack the lower stem after emergence. The black spores may be spread by wind.

If peanuts are not cured properly, they will show a high percentage of seed rot and crown rot in the succeeding crop.

The disease may develop if peanuts are planted too deeply, especially during cool, damp weather immediately after seeding.

Control

1. Treat the shelled seed with captan (4 oz/100 lb) (1 tsp/lb) immediately before planting. Add the fungicide to the seed in a plastic bag and shake thoroughly until every seed is coated.

References: 24, 64.

DIPLODIA BLIGHT

Botryodiplodia theobromae

Fig. 204. Peanut seedlings attacked at an early stage by Diplodia blight.

Diplodia blight is common on the poorer soils of Fiji. It can easily be mistaken for crown rot. It can wipe out an entire crop during the early stages of growth.

Symptoms
Stem tissues at the soil level are blackened and shred slightly. The stems may rot through completely below the cotyledons, or the plants wilt and die (Fig. 204). Diplodia blight can be distinguished from crown rot by the absence of tufts of black spores, and the presence of many black raised dots on rotted stem surfaces.

Disease Cycle
The pathogen of Diplodia blight is called *Diplodia natalensis* by some pathologists, and by others, *Botryodiplodia theobromae.* It is a very common soilborne fungus in tropical soils, and it may also be seedborne on peanuts. It attacks plants only when they are injured or weakened by some environmental condition. The most common factor appears to be lack of soil fertility. Diplodia blight has been found on highly acid red soils and silty loams.

Control
1. Build up soil fertility by rotation with grasses and other legumes, and by application of the correct fertilizer.

References: 2, 64.

LEAF ROT

Corticium sasakii

Leaf rot of peanut occurs under the same conditions as leaf rot of cabbage. The disease is not serious.

Symptoms
Leaf surfaces show large blackened shredded areas (Fig. 205). Dead leaves are stuck together by a white growth that looks like the web of a leaf-rotting insect.

Fig. 205. Severe leaf rot of peanut.

Disease Cycle
The fungus causing leaf rot is the same as that associated with leaf rot of cabbage and sheath rot of rice. The conditions favouring disease development are also the same.

Control
See control measures for leaf rot of cabbage, etc.

LEAF SPOT

Mycosphaerella arachidis, Mycosphaerella berkeleyi

Leaf spot may defoliate peanuts grown on poor soils, such as the talasiga zone of Vanua Levu and very sandy soils along the river valleys in Viti Levu.

Fig. 206. Leaf spot on the variety Local Spanish.

Symptoms

Leaves are covered with circular light brown spots that later turn reddish-brown to black and enlarge to a diameter of about ¼ inch (Fig. 206). The spots may have yellow haloes around them. In severe attacks the lower leaves die and yield may be considerably reduced.

Disease Cycle

Two fungi, that may be often present on the same leaf, cause leaf spot. One species attacks the plants at an earlier stage than the other. For all practical purposes, the spots they cause constitute one and the same disease. It is spread by spores *(conidia* produced on crop residues and on the spots on the leaves. The fungi may also be seedborne. The best conditions for disease development occur during wet warm weather.

The Local Spanish variety is considered more susceptible than Valencia or Mani Pintar.

Control

1. Check the soil fertility.
2. Grow a resistant variety.

References: 24, 64.

MARGINAL YELLOWING

Potash Deficiency

This disorder was seen recently in the Rewa Valley of Fiji but it is probably very common on the poorer soils.

Symptoms
Leaves show marginal yellowing very soon after emergence of seedlings (Fig. 207). Scorching may develop after maturity. The effect on yield is unknown. Leaf spot may be more severe on plants suffering from marginal yellowing.

Fig 207. Yellowing of the variety Local Spanish.

Cause
The symptoms are practically identical to those described and illustrated for potassium deficiency elsewhere in peanuts and other legumes, e.g., *Pueraria phaseoloides, Centrosema pubescens,* and *Calopogonium mucunoides.* This deficiency is common on soils that are highly acid or alkaline.

Control
1. Check the soil fertility.

References: 72.

BASE ROT

Ceratocystis paradoxa

New plantings suffer from this disease everywhere. It is more frequent during the wet season.

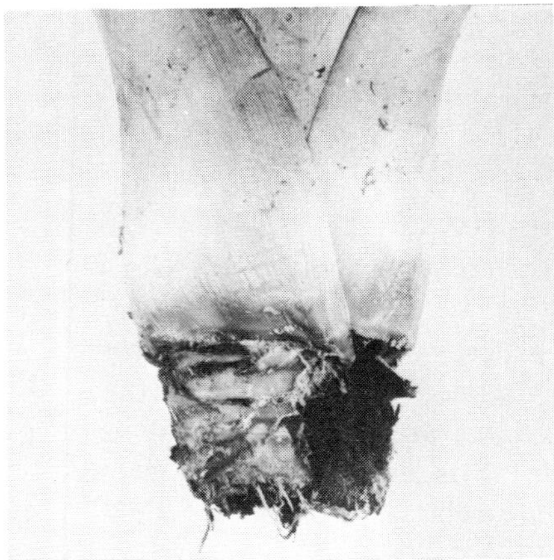

Fig. 208. Base rot on pineapple slip.

Symptoms

The cut bases of pineapple slips develop a soft brown rot (Fig. 208). It may extend upward into the leaves before or after the plants are set out in the field. If the ripe fruit is invaded it develops a yellow rot that turns black, usually starting from the cut end of the stalk. The rot has a characteristic odour. The leaves may show yellowish-white spots or broad short streaks. Older plants can be attacked through wounds at soil level.

Disease Cycle

The base rot fungus is very common in soil. It attacks healthy tissue only through wounds. When pineapple slips are not cured properly before planting, the fungus gets in during periods of rain. Spores are produced on the rotted surface and they can be spread by wind and rain. The spores of the fungus are probably present at all times on plant surfaces. If slips are stored or packed with very little ventilation without being cured, they can suffer considerable losses.

Control

1. Cure slips by turning them upside down on the tops of pineapple plants in the field for a couple of days before planting.
2. Choose dry sunny weather for planting.
3. Treat cut slips by dipping in Lanetan or Aretan for 10 minutes. Use 1 lb in 20 gallons of water. Use this method if weather is unfavourable.
4. Avoid injuring ripe fruit and the bases of mature plants.

References: 20, 36.

FASCIATION

Unknown

Fig. 209. Fasciation of pineapple.

Fasciation occurs in Rough and Cayenne pineapples more frequently than in the Queen variety. It appears only at certain times of the year.

Symptoms
More than one pineapple is formed on one stem. The crown may be laterally flattened and consists of a number of slips (Fig. 209). In severe cases, the fruit itself is flattened and monstrous.

Cause
The exact cause of fasciation is unknown. Pineapples growing in very fertile soil show more fasciation than those in poor soil. Where growth is retarded by periods of drought, fasciation may occur when favourable weather recurs around flowering time. The condition is hereditary to some extent, because fasciating clones can be selected.

The application of too much ANA to induce out-of-season flowering may also be a cause.

Control
No special measures can be recommended.

References: 20.

FRUITLET CORE ROT

Cladosporium, Curvularia, Fusarium, Penicillium spp.

Core rot is seen during harvest in very wet weather. Because it is difficult to detect externally, it causes considerable embarrassment to the grower after the fruit has been taken home from the market and cut.

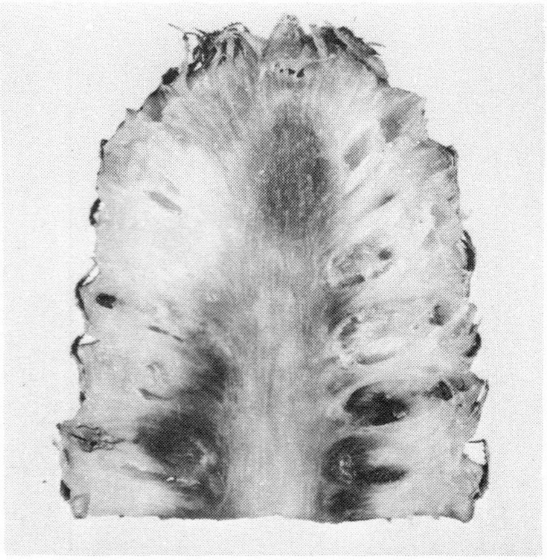

Fig. 210. An advanced case of fruitlet core rot.

Symptoms
When a ripe fruit is cut, a reddish-brown to black rot (Fig. 210) is found extending from the core of a fruitlet into the centre of the fruit. The rot is quite firm and it lacks a distinctive odour.

Disease Cycle
Weakly parasitic fungi become established on dead flower parts in the fruitlets. They may invade the carpels through or along nectary ducts that fail to close after the blossoms have fallen. After the fruit ripens, the fungal infection flares up inside the fruit. The fungi involved are soilborne and they may be blown into the fruitlets on dust or they may be carried there during hilling operations.

Control
1. Do not throw soil into the centres of plants while the fruits are ripening.
2. Pick fruits before they are too ripe.

References: 20, 59.

Phytophthora nicotianae var. *parasitica, Phytophthora cinnamomi*

Fig. 211. Heart rot in leaf bases.

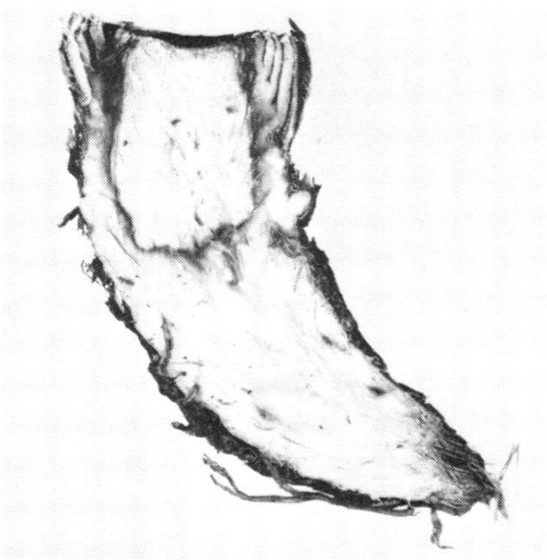

Fig. 212. Heart rot in stem.

The importance of heart and root rot has not been fully assessed because it can be easily confused with base rot in young winter plantings. It has likely caused serious losses.

Symptoms
Roots, stems, and leaf bases rot (Fig. 212). Plants turn yellowish-green and the central leaves pull out easily. The bases of affected leaves have a yellowish-white rotted area bordered by a brown margin where healthy tissue begins (Fig. 211). Young fruits may also rot.

Disease Cycle
The cycle is essentially the same as that for corm rot of dalo and root rot of avocado pear. The disease is favoured by poorly drained soil and cool wet weather. Both species of *Phytophthora* are common in tropical soils.

Control
1. Avoid poorly drained planting sites.
2. Plant during dry, sunny weather if possible.

References: 20, 36.

WHITE STREAK

Unknown

White streaking has been observed on pineapples growing at Navuso, Koronivia, and Seaqaqa Farm in Fiji. It is not important.

Symptoms
The interveinal tissue turns white commencing at the tips of the leaves (Fig. 213). It may extend the length of the leaves. It may eventually die out to form black streaks.

Fig. 213. White streak on Cayenne pineapple.

Cause
This is not known. A zinc deficiency may be inferred as the cause, because citrus in the immediate neighbourhood show excellent symptoms of little leaf. Where a complete fertilizer has been applied early in the growth cycle, white streak is not apparent.

Control

1. Where symptoms are very marked, check the soil pH and try to correct the condition by liming, according to recommendations of the Soil Chemist.

Pseudomonas solanacearum

Fig. 214. Wilted potato plant.

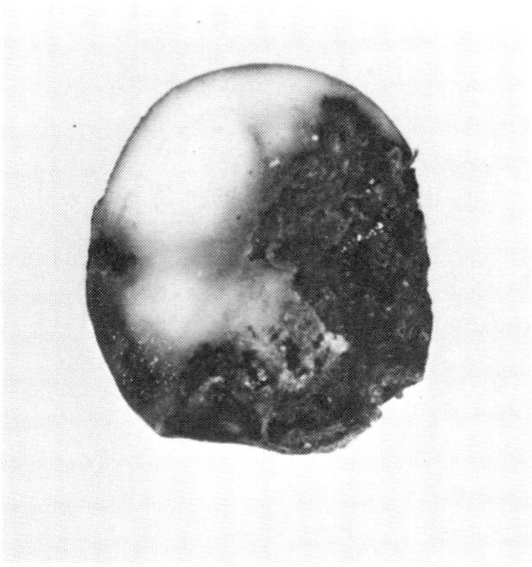

Fig. 215. Brown rot that started in the vascular ring.

Bacterial wilt is unquestionably the worst disease of potatoes in Fiji, Samoa, and Tonga. In certain places in Fiji, it has wiped out an entire crop. Some areas, however, seem to have remained free of the disease.

Symptoms

The tops wilt (Fig. 214) usually after blooming. If the stems are cut, they show a discoloration of the vascular tissue. If the bases of the petioles are cut and squeezed, a slimy grey ooze comes out of the vascular ring. Tubers show sunken black areas at the stem end. If the tubers are cut, a brown rot (Fig. 215) and grey to yellowish creamy ooze can be seen in the vascular ring. Eventually, tubers break down completely with soft rot.

Disease Cycle

The wilt bacteria have a very wide host range including tomato, eggplant, chilli, tobacco, peanut, and cowpea. It is also harboured in Fiji by the prickly solanum, *Solanum torvum,* and other solanaceous weeds.

The bacteria are motile and spread is by flooding or movement of contaminated soil or seed. If the seedpieces are cut, the cutting knife can also transmit the bacteria from infected sets to healthy ones. This means is not important locally, because most seed is imported from wilt-free areas.

Best conditions for wilt development are soil temperatures above 80°F. and dry weather.

Control

1. Avoid land infested by prickly solanum for potato culture.
2. Avoid planting potatoes too late in the cool season to avoid rising temperatures at the time of tuber formation.
3. Rotate potatoes with rice, maize, and sorghum and avoid susceptible solanaceous crops and peanuts.

References: 42, 68.

DRY ROT and SEED-PIECE ROT

Fusarium sp.

Imported seed potatoes often arrive in the islands suffering from dry rot. If they get wet during transit or after arrival considerable losses may occur. Cut seed often rots in the ground. Dry rot may also occur after harvest but it is unlikely to be serious if the potatoes are consumed immediately or if a cool store is available.

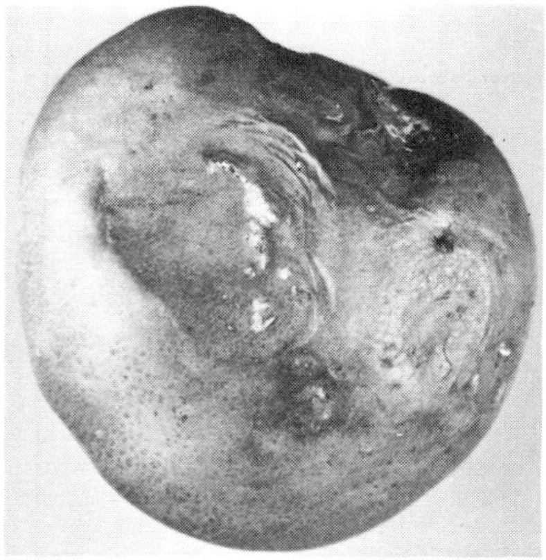

Fig. 216. Potato with Fusarium dry rot.

Symptoms

The skin of the tuber sinks and folds (Fig. 216), often in concentric circles. The whole tuber eventually becomes dry, hard, and shrivelled. The flesh is brown or bluish and the cavities are filled with a white or pink fungus growth. Often lumps of fungal growth about ⅛ inch in diameter are visible on the surface of tubers. If affected tubers are planted, they rot in the ground. An initial dry rot infection is often followed by soft rot.

Disease Cycle

The dry rot fungus is soil- and tuber-borne. It attacks tubers through cuts, bruises, and lenticels. Tubers cut before planting are quite susceptible. *Conidia* produced in and on the tubers and on undecayed organic matter in the soil spread the disease. During very dry periods and between crops the fungus forms resistant spores *(chlamydospores)* capable of resisting unfavourable conditions.

Control

1. Do not cut seed potatoes if it is possible to get smaller ones.
2. Treat cut seed with a 5% thiram dust (8 oz/bushel) (12 oz/cwt).

References: 65.

EARLY BLIGHT

Alternaria solani

Early blight is common on potatoes grown in the Sigatoka Valley, Nadala-Navai, and the Nausori Highlands. It probably looks worse than it really is, because it attacks the leaves when the plants are almost mature. It may aid tuber ripening by gradually defoliating the plants at the right time.

Symptoms
Dark brown spots consisting of concentric rings of grey and black (Fig. 217) appear on the leaves. Tubers develop symptoms similar to those on the leaves under certain conditions. The flesh of the tubers may develop pockets of dry, black rot.

Fig. 217. Target spots of early blight.

Disease Cycle
Conidia produced on long brown *conidiophores* scattered sparsely over the target spots on the leaves are responsible for spread of the disease. Under Fiji conditions of almost equal day and night during the cool season, the potato crop sets tubers very soon after the crop flowers. Foliage production is minimal and diseases such as early blight, that are associated with senescence, get an early start.

Other conditions that favour early blight are high temperatures and an imbalance between nitrogen and potash.

Control
1. Correct any imbalance in soil fertility.
2. Spray with maneb (Dithane M-22) at weekly intervals, commencing at flowering time and continuing until just before harvest, if the crop is of sufficient value.

References: 3, 7, 65.

Immaturity

This disorder is seen frequently on both local and imported potatoes. It can cause considerable losses in transit if potatoes are not properly refrigerated.

Fig. 218. Feathering on locally grown Sequoia.

Symptoms

The skin peels away (Fig. 218) easily from the flesh and curls and hangs loosely. The exposed areas may develop a secondary dry or soft rot. Tubers do not store well at room temperature and shelf life is reduced.

Cause

The cause of feathering is immaturity. Potatoes require a period of time in the ground after the vines die and dry off to harden their skins. This occurs best at soil temperatures between 60 and 70°F. A certain amount of curing will occur if tubers are left on top of the ground in cool dry weather for a day after digging. Tubers dug before the vines die have very tender skins and are most susceptible to feathering. Weak fungi and bacteria often invade feathered tubers.

Control

1. Harvest during weather favourable for proper curing.
2. Do not dig tubers until the tops have died.
3. Place tubers in a cool store as soon as possible after harvest.

SCAB

Streptomyces scabies

Scab was first observed in Fiji at Nadala-Navai in 1965 but it has probably been in the South Pacific for years. Under local conditions, it is not a serious disease.

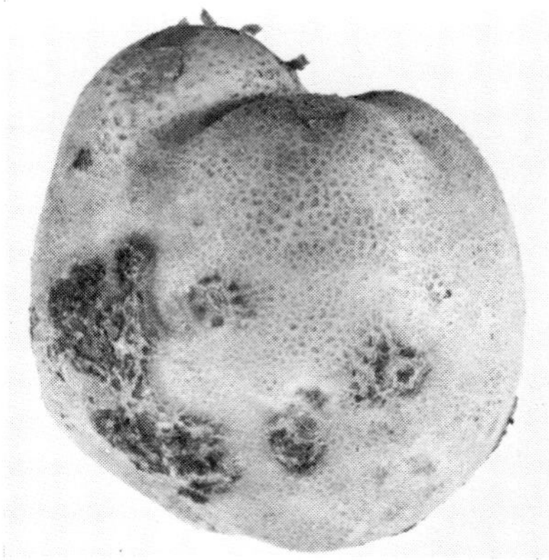

Fig. 219. Scab on the variety Sequoia.

Symptoms

Scab begins as a small brown spot around a lenticel or breathing pore. It spreads in a superficial manner and it sometimes forms craters (Fig. 219). Raised scab that looks something like human warts also occurs. The surface is very corky. Tubers are quite edible but they do not look attractive and there is considerable waste when they are peeled.

Disease Cycle

The organism causing scab is soil and tuber-borne. It invades the skin of the young tubers through lenticels and grows under the surface cells, inciting them to a corky defence reaction. The scab organism then penetrates again and the process is repeated until a visible scab is formed. *Streptomyces* spreads by breaking up the threads it produces on scab surfaces. The segments are like small bacteria. They spread through the soil by water and through the air by wind.

The scab organism definitely prefers somewhat alkaline soils and temperatures between 77° and 86°F., and dry soil conditions. It appears to be worse in low fertility soils.

Control

No special control measures are recommended at the present time.

References: 7, 65.

SOFT ROT

Erwinia caratovora, Erwinia aroideae

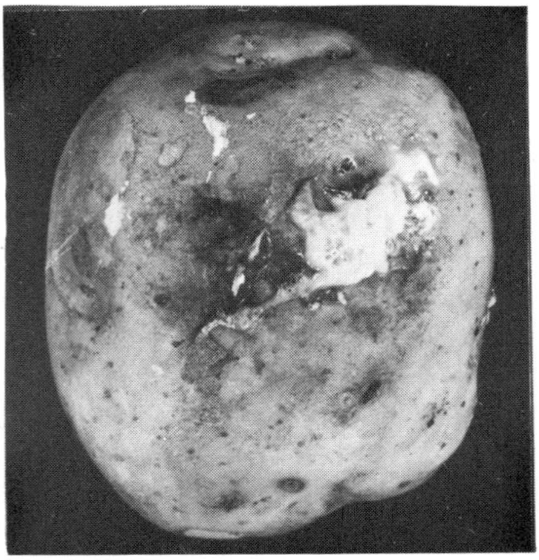

Fig. 220. External symptoms of soft rot of potato.

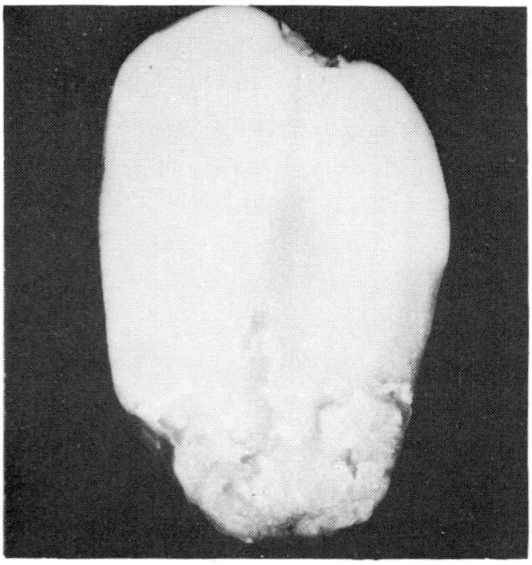

Fig. 221. Internal symptoms of soft rot of potato.

Soft rot can cause considerable losses in potatoes harvested under unfavourable weather conditions. It also invades the lower stems of a growing crop. The name suggested for this symptom is "Tropical Black Leg". If imported potatoes get wet during transit or after arrival in the islands, they may be attacked by soft rot.

Symptoms
Slightly raised spots develop around the lenticels. Cracks exuding slimy ooze (Fig. 220) develop in the skins. The tuber tissue becomes watersoaked, soft, and mushy (Fig. 221). At first there is very little discoloration but later the flesh turns brown. The rot has a very offensive odour. When soft rot from a rotting seedpiece invades the base of the stem, the affected area becomes watersoaked, collapses, and the plant wilts. The vascular system is not invaded as is the case with bacterial wilt.

Disease Cycle
The soft-rot bacteria are soilborne and they are present at all times on tuber surfaces. They get into tubers through lenticels, wounds, or after other pests and diseases. The best conditions for soft rot development are immaturity, wetting, lack of ventilation during transit or storage, and temperatures between 80 and 90°F.

Control

1. Do not cut seed if it is possible to get tubers of the right size.
2. Make sure that potatoes are mature before lifting. Look for a well-callused skin without feathering.
3. Allow potatoes to cure in the sun for at least half a day before bagging, or crating.

4. Store in trays with slatted bottoms under dry ventilated conditions at temperatures between 65 and 70°F., if potatoes are to be consumed within a few weeks of harvest.
5. If potatoes are to be kept for any great length of time, put in a cool store at 38-40°F.

References: 25, 65, 68.

WET ROT

Ceratocystis paradoxa

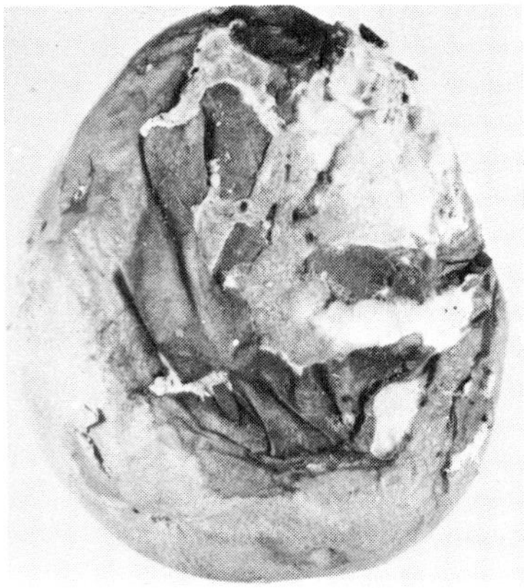

Fig. 222. External symptoms of wet rot.

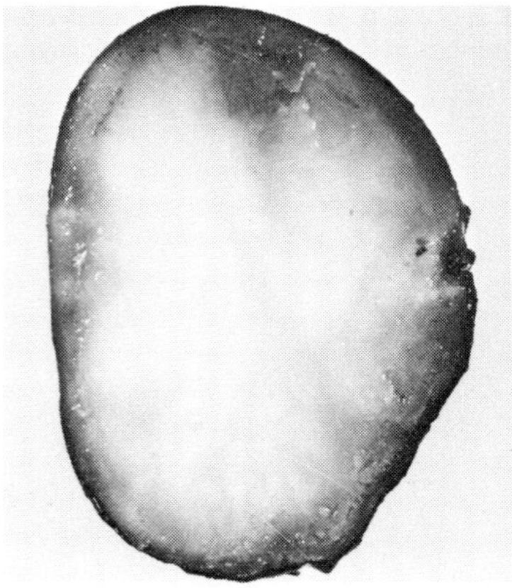

Fig. 223. Internal symptoms of wet rot.

If the weather at harvest time is very unfavourable, wet rot can develop before the crop is consumed. It may be more prevalent where copra sacks are used to transport tubers.

Symptoms

Tubers exude purple or black liquid (Fig. 222). The cut surfaces are rubbery, watery and reddish-brown (Fig. 223). A green fungous growth is evident in the cavities. Soft rot may follow wet rot.

Disease Cycle

The wet rot fungus is *Ceratocystis paradoxa,* which also causes bleeding disease of coconuts, "pineapple disease" of sugar cane setts, and butt or base rot of pineapples. It is common in soils with a lot of decaying organic matter. It cannot penetrate healthy tissues unless wounds are present.

Control

1. Do not harvest, if at all possible, during wet warm weather. Plant the crop so that it matures before the wet season.
2. Avoid the use of copra sacks for transport.
3. Avoid rough handling.
4. Put tubers to be stored for any great length of time in cool store at 38-40°F.

LEAF SPOT

Cercospora canescens, Cercospora cruenta

Leaf spot is very common on pulses during the cool season but it is doubtful that it does very much damage. It may often be confused with powdery mildew, which can be present simultaneously.

Fig. 224. Leaf spot on mung *(Phaseolus aureus).*

Symptoms

The spots develop as brown blotches, distinct from the purple markings of powdery mildew. On mung the centres of the spots turn grey or white (Fig. 224). On cowpea and long bean they may merge and if the infection is a severe one, leaves may drop. Spots may also appear on the petioles and pods.

Disease Cycle

Two species of *Cercospora* are associated with leaf spots of pulses in Fiji. The life cycles are basically the same. *Conidia* are produced on cushion-like structures *(stromata)* on both surfaces of the leaves. They are washed off during rain and splashed to healthy leaves, or they are carried there by animals, tools, or clothing. The fungi may also attack a number of wild plants in the legume family. New infections are evident in 4-8 days.

Best conditions for disease development appear to be cool weather (68-75°F.), persistent dews, and low fertility.

Control

1. Check the soil fertility
2. Apply a spray of maneb (1 oz/3 gallons) if the attack is severe and if the crop is sufficiently valuable to make the treatment economic.

References: 13, 15, 56, 69.

MOSAIC

Cucumber Mosaic Virus, Cowpea Mosaic Virus

Mosaic of several types is very common in cowpeas and long beans, but not so common in mung and urd. It is not known to reduce yields significantly, but it may be a threat to other crops, such as bananas, passion fruit, Chinese cabbage, and cucurbits.

Fig. 225. Mosaic in cowpea *(Vigna sinensis)*.

Symptoms
Leaves show a mosaic of light and dark green areas (Fig. 225), or of yellow and green. Young leaves may be severely distorted and undersized, due to irregular growth of affected areas.

Disease Cycle
Mosaic may be caused by at least 2 different viruses, cowpea mosaic and cucumber mosaic. Cowpea mosaic may be seedborne to some extent. Cucumber mosaic, with many weed hosts, may be transmitted to cowpeas by *Aphis gossypii.* It has also been reported to be seedborne in cowpea. Cultivating machinery may also spread the virus by infected plant juice.

Control
1. Insist upon a mosaic-free source of seed and if possible obtain resistant varieties.
2. Keep all weeds under control.

References: 15, 24, 74.

POWDERY MILDEW

Erysiphe polygoni

Powdery mildew is the most serious disease of cowpeas, mung, urd, and long beans during the cool, dry season. It can cause considerable loss of leaves.

Fig. 226. Powdery mildew on urd *(Phaseolus mungo)*.

Symptoms

The disease begins as a slight darkening of the foliage. Then it turns purple in irregular patches. Both sides of the leaves may become covered with a powdery white mould (Fig. 226). The older leaves shrivel, turn black, and fall off. The pods may turn purple. Yield is reduced.

Disease Cycle

The mycelium of the powdery mildew fungus invades the epidermal cells by means of *haustoria* or sucking organs. After the fungus becomes established, it pushes up chains of *conidia* that are detached from the leaf surface and blow about to infect healthy leaves. The conidia are capable of germinating on leaves *without* water droplets and in the presence of a relative humidity slightly below 100%. New mildew growth is usually evident within 8-10 days. In the South Pacific the fungus probably lives continuously in the conidial stage.

Conditions for disease development are apparently best during the cool, dry season, when air temperatures are between 68° and 75°F.

Control

1. Plough under diseased refuse from pulse crops, and rotate them with other mildew-resistant vegetables, e.g., cabbage, tomato, eggplant.
2. Spray with Karathane (2 oz/3 gallons) if the crop is sufficiently valuable. Two weekly applications after mildew appears should be adequate.

References: 2, 15, 24, 62, 69.

Uromyces appendiculatus, Uromyces vignae

Fig. 227. *Left:* Rust pustules on runner bean pod; *right:* rust on long bean leaflet.

Rust has been observed on runner bean *(Phaseolus coccineus)* and on long beans *(Vigna sesquipedalis)* at Koronivia, and on haricot beans and various pulses on cane farms in the Ba district, in Fiji. It is of minor importance.

Symptoms

The disease commences as small, whitish, slightly raised pustules. They enlarge to reddish-brown, cushion-like spots about $^1/_{16}$-$^1/_8$ inch in diameter, with yellow haloes (Fig. 227). Sometimes the spots are surrounded by outer rings of pustules. They may be so numerous that leaf surfaces are completely covered and obscured. Severely infested leaves turn yellow, dry up, and fall off. Pustules may also appear on the pods (Fig. 227).

Disease Cycle

The rust species on long bean is apparently different from that on other species of pulses. Therefore haricot beans may appear quite healthy when grown close to long bean, and *vice versa.*

Only the *uredospores* spread the disease locally. In colder climates, *teliospores* carry the disease through the winter and then germinate to form *basidiospores* that recommence the disease. The *uredospores* are produced in the reddish-brown pustules in great numbers. They shake out of the pustules and blow to other leaves, where they germinate in dew or raindrops and send their germ tubes into the stomata. Symptoms appear under favourable conditions (60-75°F.) within 5 days of infections, and fresh uredospores are produced 5 to 10 days later.

Cloudy humid weather and persistent dews favour disease development.

Control

1. Use a long crop rotation.
2. Spray with zineb (1 oz/3 gallons).
3. Experiment with resistant varieties. In the case of an outbreak of rust on haricot beans, Hawaiian Wonder, Lualualei, Seminole, Criolla, Florida Belle, Dixie Belle, and Fullgreen, may be worth while.

References: 6, 14, 15, 70.

BLAST

Pyricularia oryzae

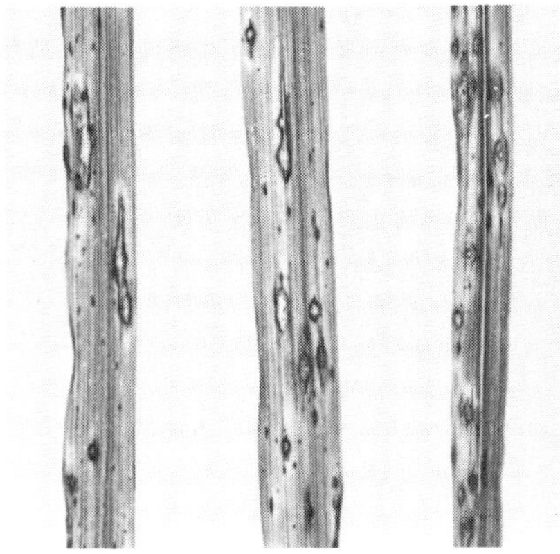

Fig. 228. Leaf blast on BG-79.

Fig. 229. *Left:* Rotten neck; *right:* healthy neck.

Although blast or "rotten neck" is a serious disease of rice elsewhere in the world, it has not been important in Fiji until now. In 1966, it looked serious on some seedling fields of BG-79, but they recovered close to harvest. Surveys have indicated that BG-79 is a blast-susceptible variety.

Symptoms

Bluish flecks appear on the leaves. The flecks enlarge to form irregular brown spots with ashy-grey centres (Fig. 228). In severe attacks, the spots run together and the whole leaf turns brown. Spots may also appear on the culms.

Major damage by blast is caused when the neck or joint just below the panicle is attacked and rots. If this happens early, the grain never fills. If it happens late, the panicle falls over (Fig. 229). A grey, fluffy fungus growth can be seen on the affected parts.

Disease Cycle

The blast fungus may be carried on seed. It is spread mainly by the bottle-shaped *conidia,* that are produced on any affected part of rice plants, even stubble that persists between crops. They may also be formed on grasses growing in and near rice fields. The *conidia* germinate on rice leaves and they send germ tubes down through the stomata. Young leaves and spikelets are most susceptible.

The best conditions for infection are high humidity and temperatures between 75 and 85°F. The disease is definitely worst on soils with a high level of nitrogen.

A large number of rice varieties and selections are resistant to blast, but breeding for resistance is complicated by strains of the blast fungus. After a time, blast-resistant varieties appear susceptible because of changes in strains.

Control

1. Do not put high nitrogen fertilizer on rice.
2. Grow resistant varieties if they are available.

3. Do not use seed from severely infested fields. Dust seed with a dry mercurial dust (2 oz/100 lb).
4. Burn stubble and straw or plough it down as soon as possible after harvest.

References: 24, 40, 60.

BROWN SPOT

Cochliobolus miyabeanus

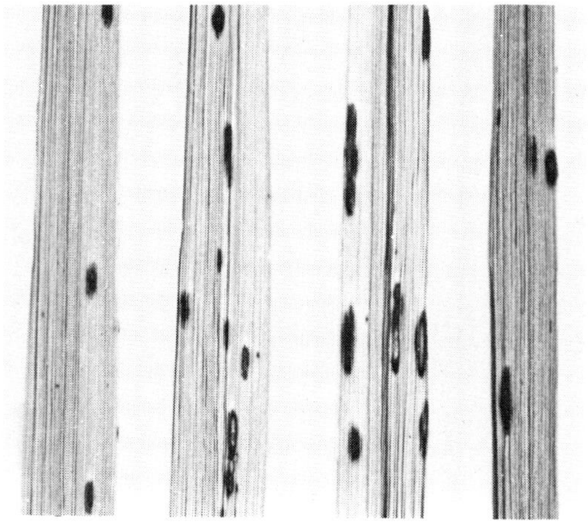

Fig. 230. Brown spot on leaves of BG-79.

Fig. 231. Brown spot on grain of BG-79.

Brown spot is not a serious foliage disease of rice in the South Pacific but it can cause a great deal of seedling blight when it attacks the grain. It is not regarded as serious as blast. Generally, dryland rice, under conditions of continuous culture and low fertility, suffers a great deal more from brown spot than wetland rice.

Symptoms

In the early stages brown spot is not easy to distinguish from blast. Spots are more regular in outline with borders (Fig. 230). It usually appears much later in the season than blast. On the grain, "eye spots" with brown edges and grey centres develop (Fig. 231). In severe attacks, the spots are covered by black spore masses. Seed may become shrivelled. When the leaf sheaths are stripped away, the culms show large brown lesions often over 1 inch long. In severe attacks plants may collapse.

Badly affected seed may rot in the ground, or seedlings may die before or after emergence. Survivors are stunted, yellowed, but not wilted as in the case of infection by the fool's rice or blast fungi.

Disease Cycle

Spots appear about 3-4 days after infection by germ tubes penetrating directly into the leaf epidermis from spores that have landed there. After about 2 weeks a fresh crop of long brown spores is produced. These *conidia* are called the Helminthosporium state. They are blown about by wind and they germinate in water droplets. Older leaves appear to be more susceptible to brown spot than younger ones. The *perithecia* of the sexual or Cochliobolus state are found occasionally on rice stubble. *Ascospores* shot out of the *perithecia* can also infect rice leaves. The fungus can also attack a number of grasses in or near rice fields.

The best conditions for brown spot are a high humidity and temperatures of 80-90°F. Seedling blight is definitely more serious early in October than in January, when high soil temperatures speed up the growth rate of rice.

Control

1. Avoid poor soils and keep up the fertility levels on 2nd and 3rd year rice.
2. Select seed from clean fields and treat with a dry mercurial dust at the rate of 2 oz/100 lb of seed.
3. Irrigate the seedbed if possible in the case of rice to be transplanted.
4. Do not sow rice land too early when soil temperatures are low.
5. Burn straw and stubble or plough it down deeply as soon as possible after harvest.

References: 24, 60.

FALSE SMUT

Ustilaginoidea virens

A very small amount of false smut can probably be found in any rice field in Fiji, but the disease rarely causes serious losses. It appears to be more common along the south coast of Viti Levu than in the river valleys or the dry zones.

Fig. 232. False smut on the variety BG-79.

Symptoms

Grains are transformed into large, velvetty, green masses over twice the size of normal grains (Fig. 232). The masses eventually burst to expose a bright orange interior. Usually only a few grains in a head are affected.

Disease Cycle

Orange-coloured spores *(conidia)* are produced all over the surface of the large masses *(sclerotia)* that have replaced the grain. The conidia germinate in water to produce secondary conidia. They probably infect the flowers before the rice heads have fully emerged from the leaf sheaths. The flower ovaries are destroyed and replaced by sclerotia. Invasion may also take place when the grain is mature.

It is doubtful that the false smut fungus is effectively seed or soilborne. It may attack grasses as primary hosts and then move on to rice. Knowledge of the life cycle and best conditions for infection is not complete.

Control

No special measures are recommended.

References: 24, 60.

FOOL'S RICE

Gibberella fujikuroi

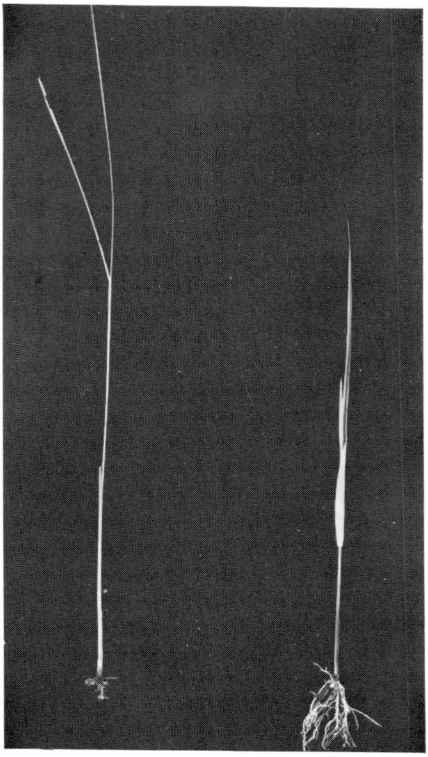

Fig. 233. *Left:* Root rot caused by a heavy soilborne infestation; *right:* healthy root system (above).

Fig. 234. *Left:* Seedling showing fool's rice or "bakanae" symptoms; *right:* healthy seedling (left).

Fool's rice, or "bakanae" disease, is the most serious seed- and soil-borne disease of dryland rice in Fiji, particularly in the Rewa Valley of Viti Levu.

Symptoms

Seed emerges poorly and gaps appear in the rows. Affected seedlings are very thin, yellow, and taller than healthy ones of equal age (Fig. 234). The sick ones eventually wilt and die, showing a pink-white dusty covering of fungus spores at soil level. Those infected at a later stage grow very tall, due to abnormal elongation of the internodes, and they do not fill their grain.

If the leaf sheaths are stripped away, the nodes on the culms are pink to purple in colour. Root rot (Fig. 233) may occur if the soil is heavily infested. Infected grain is pink in colour.

Disease Cycle

The fool's rice fungus may be seed or soilborne. Infection of seed occurs at flowering time, when the very small airborne *conidia* of the Fusarium or asexual state germinate and invade the glumes of the developing grain. If seeds are lightly contaminated, the seedlings show the "bakanae" effect. If they are heavily infested, usually from a soilborne source, they may blight or develop root rot. The fungus can grow up from underneath the soil line on to the culms and leaf sheaths. *Conidia* are produced abundantly on straw or stubble and they blow easily because they are very light. The fungus also produces *perithecia, asci,* and *ascospores* under certain conditions. The ascospores do not appear to be very important in spreading the disease locally.

The fool's rice fungus can also attack maize, sorghum, sugarcane, and many grasses. Not all strains will produce the "bakanae" symptoms, which is a reaction peculiar to rice.

Best conditions for development of "bakanae" are high soil temperatures (90-100°F) and dryland culture. Lower soil temperatures and very dry conditions are more suitable for root rot and stunting.

Control

1. Treat seed with a dry mercurial dust (2 oz/100 lb).
2. Avoid a high-nitrogen fertilizer.
3. Rotate rice with non-susceptible crops such as vegetables or root crops after the second year.
4. Burn straw and stubble or plough it down as soon as possible after harvest.

References: 24, 60.

FOOT ROT

Ophiobolus oryzinus

Fig. 235. *Left:* Rotted culm split down the middle; *centre:* rice affected by foot rot; *right:* healthy plant.

Foot rot, or brown sheath rot, was found at Koronivia in Fiji in 1964. It is of very minor importance but it is interesting because it has a limited distribution in the U.S.A., one African country, the Philippine Islands, and Papua-New Guinea.

Symptoms

Just before harvest the bases of the leaf sheaths for about 6 inches above the soil become shredded and greyish-brown (Fig. 235). If the sheaths are peeled away the culms show a brown rot often covered by a brown fungus growth. If the outside of the sheaths are examined closely, black pinpoints can be seen protruding from their surfaces. Plants severely affected by foot rot collapse at soil level.

Disease Cycle

The foot rot fungus produces only *perithecia,* or black pinpoints, on infested leaf sheaths and stubble. The *ascospores* that are tightly coiled in *asci* inside the perithecia are shot out at maturity and carried by irrigation or rain water to healthy leaf sheaths. Their germ tubes penetrate the sheaths directly through the epidermis and then the hyphae of the fungus spreads beneath them on to the surface of the culms, rotting them through. Then perithecia develop in the rotten tissue and push out through it.

Best conditions for development of foot rot appear to be close planting and temporary flooding of plants as they approach maturity.

Control

1. Burn stubble and straw or plough down to a depth of 9-10 inches after harvest.

References: 4, 24, 60.

GREY LEAF SPOT

Pyrenochaeta oryzae

Grey leaf spot occurred in Fiji in 1965 on rice leaves in an introduction nursery. Very little damage was done as the plants were close to harvest.

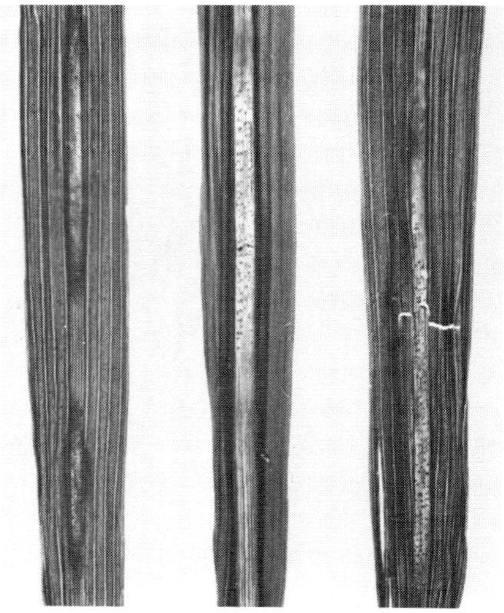

Fig. 236. Grey leaf spot on the variety Taichung Native No. 1.

Symptoms

Elongation rusty brown spots, turning grey (Fig. 236) extend for 1 to 2 inches along the midveins of rice leaves. Sometimes the spot cracks in the centre and the leaf droops away from the culm.

Disease Cycle

The fungus *Pyrenochaeta oryzae,* a weak parasite, was found consistently associated with grey leaf spot. It apparently invades the feeding or egglaying wounds of leafhoppers.

It was identified on the variety Taichung Native No. 1 introduced from the Philippines. It is most unlikely that it was carried into Fiji with this variety, because it is a very common fungus on seed of rice and other grasses.

Control

No special measures are recommended.

References: 60.

NARROW BROWN LEAF SPOT

Cercospora oryzae

This disease is not seen often on rice varieties now grown commercially in Fiji. It has been observed only in variety trials at Koronivia.

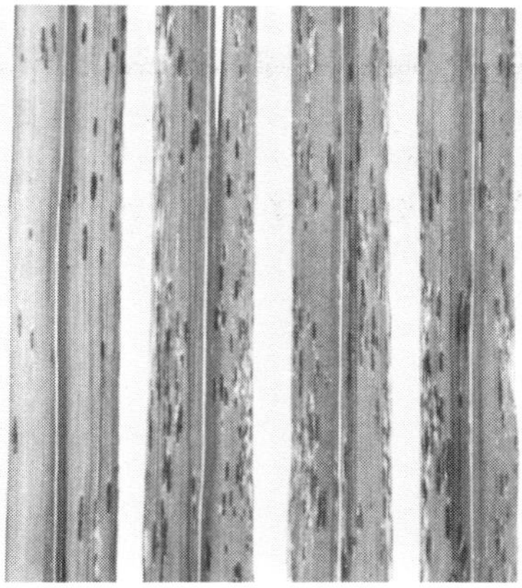

Fig. 237. Narrow brown leaf spot on Koronivia Patarka.

Symptoms

Spots are narrow and linear (Fig. 237), somewhat like those of Sigatoka disease of bananas. They are reddish-brown, $\frac{1}{4}$-$\frac{1}{2}$ inch long, and with light brown margins. Spots can also appear on the sheaths, culms, and flower parts.

Disease Cycle

The leaf spot fungus, a species of *Cercospora* similar to those causing Sigatoka disease and black leaf streak, produces large numbers of *conidia* on *conidiophores* on the surfaces of spots. These are blown by wind to healthy leaves, where they germinate in water droplets and grow down through the stomata. The mycelium invades the cells just underneath the stomata, and spreads up and down the leaf. *Conidiophores* are then produced up through the stomata, and a new crop of conidia is formed.

Best conditions for disease development are not known. The fungus is known to attack certain varieties more severely than others. Koronivia Patarka is susceptible, but the popular BG-79 is resistant. Elsewhere, Rexoro, Texas Patna, and Bluebonnet 50 are susceptible. About 10 races of the pathogen have been distinguished in the U.S.A.

Control

1. Grow resistant varieties.

References: 4, 13, 24, 60.

SHEATH ROT

Corticium sasakii

This disease is not important in transplanted and wetland rice. It can be serious in fields of dryland rice that are drilled too densely or where the rice is growing very lushly.

Fig. 238. Severe sheath rot.

Symptoms

The lower leaf sheaths and sometimes the leaves show large grey patches with dark brown or black edges (Fig. 238). They may be 1-3 inches long. A cobwebby growth of fungus threads can be seen on the rotted areas. Oval, greenish-brown cushion-like bodies *(sclerotia)* about half the size of a pea are often seen on the spots. If the plants are severely affected, they wither and die.

Disease Cycle

The cushion-like bodies or sclerotia spread the disease. They are very resistant and they can persist in the soil for years in a dried state. They are carried about in flood water and in soil that is stuck to machinery. When the sclerotia come into contact with rice plants, they germinate and the fungus threads from them grow over leaf sheaths, putting down infection pegs here and there. Best conditions for growth are high moisture and high temperature. These occur in dense stands.

Control

1. Avoid too close a spacing in drilled rice.
2. Avoid too high a level of nitrogenous fertilizer.
3. Burn badly infested stubble and plough it under a depth of 9-10 inches.

References: 60.

STEM ROT

Leptosphaeria salvinii

Fig. 239. *Left:* Rotted culm split to show sclerotia; *centre:* rice affected by stem rot; *right:* healthy plant.

Stem rot was first reported in Fiji about 1955. It appears to be closely associated with lodging.

Symptoms

The most common symptom is late tillering when the crop should be ripening. Ears are light and the grain is chalky. If the culms are split, a dark grey fungus growth with small, round black pinpoints (Fig. 239) can be seen in the hollow part and on the inner surfaces. The attack begins at the first node or higher.

Disease Cycle

The black pinpoints *(sclerotia)* spread the disease. Under certain conditions the fungus may also produce *perithecia, asci, ascospores,* and *conidia* resembling those of *Helminthosporium.* Sclerotia from stubble and straw float in irrigation water to young rice plants. They germinate to put out *hyphae* that penetrate unwound tissues. Sclerotia can live for as long as six years in soil.

Best conditions for infection are wet land and rice plants weakened by a fertilizer imbalance, e.g., excess nitrogen and not enough potash.

Control

1. Burn straw and stubble, if possible, just after harvest.
2. Apply high-potash fertilizer in accordance with results of soil tests.

References: 4, 24, 60.

STRAIGHTHEAD

Nonspecific

A condition resembling straighthead has been observed in wetland rice trials at Koronivia in Fiji. Certain varieties are apparently susceptible. Its effect on yield under local conditions has not been measured.

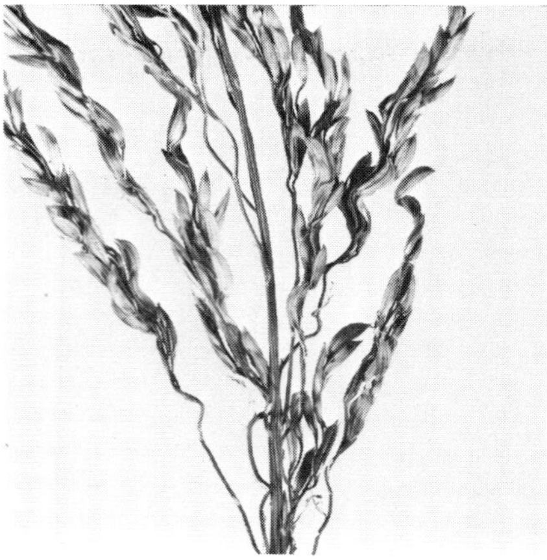

Fig. 240. Curved empty glumes on panicle of Patarka rice affected by straighthead.

Symptoms

Rice panicles remains upright at maturity. The glumes are curved like a beak and also the racemes may be twisted (Fig. 240). Often one or both glumes may be missing. In severe cases the panicles are smaller than normal and they emerge slowly from the boot, or not at all. Leaves are dark green and stiff. The roots tend to be large, without branches and root hairs.

Cause

Straighthead is attributed in the U.S.A. to an abnormal soil condition associated with prolonged flooding. If the soil contains too much undecayed organic matter from a previous crop that has been ploughed down, straighthead may develop. The varieties Rexoro and Sunbonnet are very susceptible, while Fortuna, Bluebonnet, and Texas Patna are resistant.

Control

1. Drain rice fields just before growth reaches the shooting stage, let the soil dry thoroughly and then reflood for the rest of the season.
2. Avoid very susceptible varieties.
3. Allow crop residues to decay before seeding time.

References: 4, 24, 60.

Phosphorus Deficiency

Fig. 241. *Left:* Phosphorus-deficient seedling: *right:* normal seedling (left).

Fig. 242. *Left:* Deficient seedling with hairy roots; *right:* a normal root system (above).

This disorder is seen occasionally in the Rewa Valley and along the south coast of Viti Levu. It has a tendency to correct itself after a flood.

Symptoms
Plants are stunted (Fig. 241), tillers are reduced in number, and the leaves are tightly bunched. In some varieties the leaf sheaths are purple instead of dark green. The roots are reduced and very hairy (Fig. 242), flattened, thickened, and purplish-brown in colour. Plants rarely produce grain that is properly filled.

Cause
Experiments conducted elsewhere have shown a very similar condition in rice grown without phosphate fertilizer. In Fiji the condition is often observed on hardpan soil and it may in fact be caused by a lack of moisture which makes phosphorus unavailable to the rice plants. It has not been observed in wetland rice. Flooding tends to correct the condition.

Recent experimental results indicate that phosphorus is very critical during the seedling stage of rice and that it is essential for normal root development. A shortage of this element is also very likely to occur in highly acid soils.

Control
1. Apply superphosphate and possibly lime at seeding time in accordance with the results of soil tests.

References: 60.

LEAF SPOT

Cercospora sesami

Leaf spot is common on sesame almost everywhere it is grown. Locally it is not considered a serious disease.

Fig. 243. Cercospora leaf spot of sesame.

Symptoms

Spots are about $\frac{1}{16}$ to $\frac{1}{8}$ inch in diameter, circular, with papery grey centres and purple margins (Fig. 243). Sometimes they run together and whole areas of the leaves turn purple, then they yellow and fall off.

Disease Cycle

The disease is spread by *conidia* that are produced in tufts on the surfaces of the spots.

It appears to be worst during the cool season. Local varieties may be susceptible.

Control

No special measures are recommended in view of the low value of the crop.

References: 13.

HEAD BLIGHT

Glomerella, Fusarium, Nigrospora, Phyllosticta spp.

Head blight is very destructive in high-rainfall zones. It is often mistaken for smut, which is not known to occur in Fiji. The disease causes poor grain quality and a low germination rate.

Fig. 244. Head blight of sorghum.

Symptoms

The heads turn black at or close to maturity (Fig. 244). If the individual grains are examined, it can be seen that they have burst exposing the white endosperm. This becomes blackened by fungus growth. The glumes are covered by brown or black spots. If seed from blighted heads is planted, a great deal of seed rot and seedling blight results.

Disease Cycle

A number of weakly parasitic fungi are normally present on dead glumes. They are also soilborne. As the grain matures, these fungi multiply. If harvest weather is very wet, their numbers are very high. Immature grain often bursts when wetted, and the fungi move in. If flooding occurs after flowering, soil containing many fungus spores is deposited on the sorghum heads.

Control

1. Grow sorghum in dry zones.
2. Treat seed with a dry mercurial seed dressing (2 oz/100 lb).

References: 67, 77.

Puccinia purpurea

Rust is the worst leaf disease of sorghum and also broom corn but its severity varies with the zone in which it is grown. It is very bad in wet zones with over 100 inches of rainfall annually, but it is only moderate in zones with 70 inches or less. The red-pigmented variety Alpha is quite susceptible but the white variety Combine Shallu is resistant.

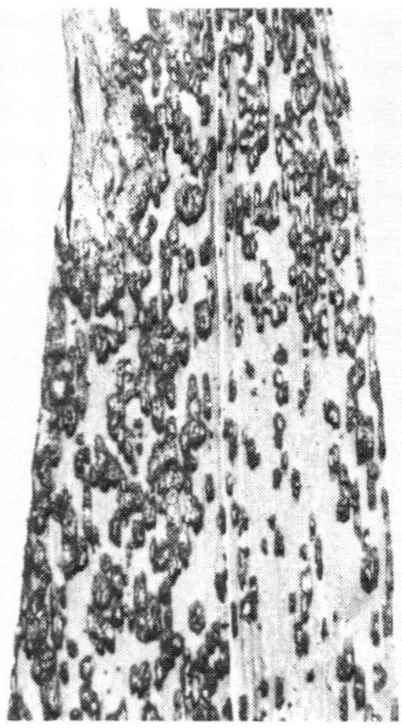

Fig. 245. Rust on the variety Alpha.

Symptoms

Purple, red, or tan flecks appear on the leaves. On resistant varieties these are the only symptoms. On susceptible varieties the spots enlarge to a length of $\frac{1}{8}$-$\frac{1}{4}$ inch and eventually break open to expose a mass of reddish-brown spores (Fig. 245). Usually there is a purple border around each pustule, but certain varieties lack this pigment.

Disease Cycle

Uredospores produced in pustules known as *uredia* spread the disease. As far as is known, the uredospore is the only spore type produced in this part of the world. In other countries, *teliospores* may be formed after cold weather sets in, because they are adapted to colder winters. The best conditions for rust development are the same as those for maize rust.

Control

1. Grow the resistant variety Combine Shallu or others that may be available from overseas.

References: 24, 77.

TOP ROT

Gibberella fujikuroi

Only two diseased plants were detected in 1966. They were found at the Agricultural Station in the Sigatoka Valley. Top rot may become more serious if the acreage of sorghum near rice and sugar cane is increased.

Fig. 246. Sorghum head choked by a rotted leaf sheath.

Symptoms

Heads fail to emerge fully from the uppermost leaf sheath, which may be twisted and covered with red streaks and a pink or white fungus growth (Fig. 246). The heads are rotted and they smell badly.

Disease Cycle

The top rot fungus also causes "pokkah boeng" of sugarcane, "fool's rice", and it is involved in pink ear rot of maize. It is a very common soilborne fungus which thrives on cereal crop refuse and on grasses. It is transmitted by the very light spores that are produced in great numbers on rotting plant parts. It may form resistant spores *(chlamydospores)* that make it capable of surviving very dry conditions, which may prevail between crops.

Infection occurs when young sorghum leaves or leaf sheaths are injured in some way. Some infection may occur directly through flower parts. Apparently the disease is worse under wet, cool conditions when sorghum is not growing fast enough to protect itself.

Control

1. Apply a dry mercurial seed dressing (2 oz/100 lb of seed).
2. Rogue and burn diseased plants as soon as they are detected.
3. Rotate sorghum with non-cereal crops.

References: 67, 77.

LEAF SPOT

Cercospora henningsii

Leaf spot is very common on tapioca everywhere in the South Pacific but it causes no significant damage.

Fig. 247. Cercospora leaf spot on leaves of tapioca ready to harvest.

Symptoms

The lower, ageing leaves show large brown blotches (Fig. 247) about ½-1 inch in diameter. Spots on the younger leaves are very small. If the attack is severe, some leaf drop may occur but this is normal for mature tapioca.

Disease Cycle

The leaf spot fungus on tapioca has a life cycle similar to that of species of *Cercospora* on kumala, cucurbits, cowpea, tobacco, and many other cultivated and wild plants.

Best conditions for leaf spot appear to be the presence of maturing foliage on the plants.

Control

No special measures are needed.

References: 13, 56.

Genetic

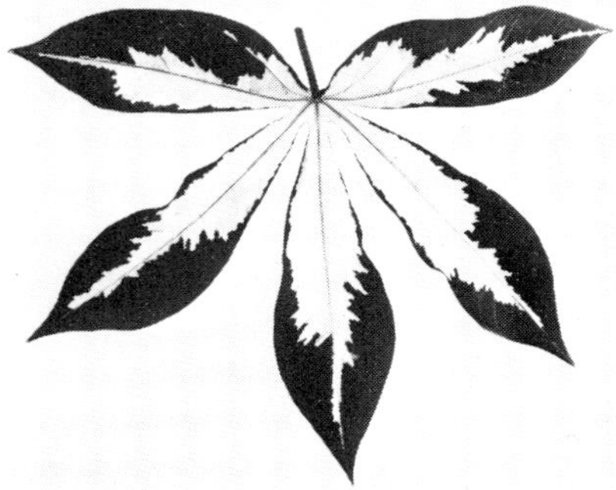

Fig. 248. Variegation in ornamental tapioca.

Fig. 249. Variegation in the tapioca variety Coci.

Variegation is uncommon in tapioca, but it causes concern when it occurs. It has no effect on yield, as far as is known.

Symptoms

Plants are somewhat stunted. The leaves are crinkly and the edges are banded by dark green, in contrast to the light green of the area along the midrib (Fig. 249).

Cause

Variegation is *not* a disease, although it vaguely resembles the virus disease mosaic, which is limited in its distribution to parts of Africa and Java. It is a characteristic of certain tapioca varieties, e.g., Coci (pronounced "thothi"). Variegated varieties have been grown in close proximity to normal green varieties for years, but the variegation does not spread. It probably originated with a *mutation,* or change in the hereditary makeup of the plant. Tapioca, like all flowering plants, sets seed which is not completely uniform. Seedlings with variegated leaves were probably propagated first by accident, and then kept as a variety. Ornamental tapioca (Fig. 248) probably originated in the same way.

Control

No control measure is necessary, if the varieties are acceptable in other ways.

Nonspecific

Fig. 250. Tea bush with early symptoms of dieback.

Fig. 251. Basal canker on tea bush shown in Fig. 250.

Dieback was observed on tea bushes about one year old in an experimental planting in the Lomaivuna Settlement Scheme in Fiji. Only 4 bushes were affected.

Symptoms

Branches in the centres of the plants lose their leaves and die (Fig. 250). The roots are apparently quite healthy. Cankers (Fig. 251) may or may not be present on the main branches and stems.

Disease Cycle

The local trouble was traced to injury to tissues that had become too woody to bend after pegging which is supposed to stimulate bush formation. Dieback is also attributed to pruning injury, delayed hurricane damage and basal cankers caused by close cultivation and weed removal.

In other countries, causes of dieback include: (1) pruning when the tea bush has depleted carbohydrate reserves, associated with high growth rates and high respiration losses, particularly at low altitudes; (2) careless pruning; (3) drought; (4) slow death of shade trees or previous forest cover, stimulating fungal diseases which attack the roots.

Control

1. Prune at the proper time of year, preferably when carbohydrate levels are highest. This can be determined only by a tea agronomist.
2. Prune close to buds, making clean cuts with sharp knives. Protect large cuts with tar or a fungicidal paste.
3. Ringbark shade trees intended for removal, or treat them with a herbicide.
4. Exercise care in cultivation. Avoid injury to roots.

References: 27.

GREY BLIGHT

Pestalotia sp. *Glomerella* sp.

Grey blight has been observed on seedling tea in an experimental plot in the Lomaivuna Settlement Scheme in Fiji. It appears to be of minor importance.

Symptoms
Large grey and brown patches with a zonate pattern develop at any point on the leaf surface (Fig. 252). The spots often crack and shrivel.

Fig. 252. Grey blight on seedling tea leaf.

Disease Cycle
The fungi associated with grey blight are weak parasites that require some kind of a wound for entry. Mechanical damage, sunscorch, and insect injuries could all be ports of entry. The disease fluctuates in severity depending upon the weather and the amount of cloud cover.

Control
No special control measures are recommended.

References: 27.

RED RUST

Cephaleuros pulvinatus

Red rust is widespread but it is not known to cause serious damage. The algal species associated with the disease in Fiji may not be the same as that in other tea producing countries.

Fig. 253. Red rust on neglected tea.

Symptoms
Orange-red raised patches, circular in shape and varying in diameter from $^1/_8$-$^1/_4$ inch, develop on leaves and branches (Fig. 253). Leaves drop and stems die back. Some cracking of bark may occur. Affected bushes are thin, weakly, and they fail to produce new growth.

Disease Cycle
The alga or pondweed associated with red rust in Fiji is *Cephaleuros pulvinatus*, which may be different from that reported on tea elsewhere, *C. mycoidea.* It is also common on avocado, cocoa, coffee, annatto, mango, and other broadleaved trees. Unless the host plants are growing under very unfavourable conditions: excessive shade, insufficient shade, lack of moisture, low fertility, the alga does not cause much trouble.

The green stalks *(sporangiophores)* of the alga bear oval *sporangia* that germinate in water to produce swimming spores *(zoospores)* that are spread by splashing rain. They eventually settle down on the leaf surface and put out germ tubes that penetrate the epidermal cells, or the lenticels of the bark.

The growths are most common during the wet season, but symptoms of injury are most severe in the subsequent dry season.

Control
1. Prune away affected branches. Stimulate the tea bush into fresh growth by the application of the appropriate fertilizer.
2. In severe cases spray with Kopi or Cuprox (2 oz/3 gallons).

References: 27.

ROOT KNOT

Meloidogyne incognita

Root knot was found in a very small percentage of tea plants in a pilot trial at the Lomaivuna Settlement Scheme, in Viti Levu, Fiji. The trouble may have commenced in nursery soil.

Fig. 254. Severe root knot of tea with overgrowth of *Sclerotium rolfsii.*

Symptoms

The plants, about a year old, showed symptoms of leaf drop, wilting, and dieback. The tap roots were transformed into large galls (Fig. 254). Lateral and fine root development was completely arrested.

Disease Cycle

The larvae of *Meloidogyne* are reported to gain entry to tea roots through root tips, and to travel from there into the more mature tissues. It is quite possible that the trouble in Fiji commenced in infested nursery soil, and was perpetuated after transplanting into the field. On the other hand, root knot is common on weeds in the field site.

The life history of the root knot nematode in tea is similar to that described for the same species on banana, carrot, tomato, tobacco, etc.

Control

1. Fumigate nursery soil with formalin or methyl bromide.
2. Eliminate highly susceptible shade trees, e.g., dadap *(Erythrina lithosperma).*
3. Before planting tea, grow a clean-up green manure crop of Guatemala grass *(Tripsacum laxum)* or of a resistant legume, e.g., *Crotalaria, Glycine, Phaseolus atropurpurea.* Marigolds *(Tagetes patula, Tagetes erecta)* may also be used because they attract the nematodes, which then die from toxins produced by the roots.
4. Treat infested sites in the field with Fumazone or EDB, after pulling and destroying infested plants. (See Soil Fumigants, page 234).

References: 27, 32.

FLECKING AND SHOTHOLE

Phosphorus Deficiency

Fig. 255. Flecking of cured leaf.

Fig. 256. Flecking and shothole on lower tobacco leaf.

This disorder has been particularly severe near Vatukarasa, Viti Levu, Fiji, but it has been seen elsewhere on rather light soils. Where it is severe the lower leaves cannot be processed.

Symptoms

Small white spots (Fig. 255) develop between the veins and around the leaf margins. They may enlarge to ⅛ inch and then run together. Then the centres drop out of the spots (Fig. 256) and the leaf may become a skeleton.

Cause

In Fiji the trouble has been traced to a low availability of phosphorus in the soil. If associated with a deficiency of potash, the shothole symptom may be accentuated. Similar symptoms have been induced by withholding phosphorus under experimental conditions. The deficiency has been accurately described in overseas references.

Control

1. Ensure an adequate supply of phosphate fertilizer to seedlings.
2. Keep a close check on fertility on light soils.

References: 37, 50.

FRENCHING

Bacillus cereus

Frenching is common everywhere tobacco is grown but it seldom causes much damage. It has been observed in Viti Levu, Fiji, on poorly drained soil in the Vatukarasa district.

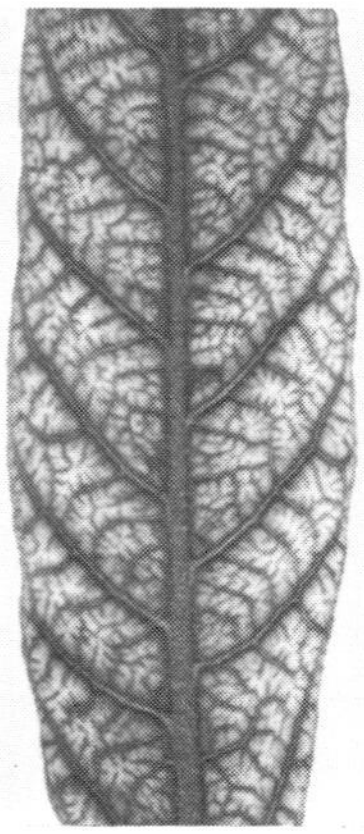

Fig. 257. Young tobacco leaf showing symptoms of frenching.

Symptoms

Margins of the young leaves turn yellow, and then the interveinal areas turn yellow too. The veins themselves remain dark green. The leaves become straplike (Fig. 257). Sucker development is excessive and affected plants are broomlike in appearance. The lower leaves of frenched plants may be yellowed.

Disease Cycle

Frenching remained a mystery for years until an American investigator demonstrated that a common soil bacterium, *Bacillus cereus,* could produce a toxin that disturbs nitrogen metabolism in the tobacco plant. Certain substances *(amino-acids)* are then apparently produced in excess and they cause the frenching symptoms. *Bacillus cereus* is a very common inhabitant of soil and plant surfaces, and it has been isolated from the root surfaces of tobacco.

Frenching may develop more readily in wet, poorly aerated soils, and also in those soils lacking in nutrients. It develops best at high soil temperatures (95°F).

Control

1. Supply adequate fertilizer.
2. Supply adequate drainage, or avoid poorly drained soils.

References: 37, 50.

Cercospora nicotianae

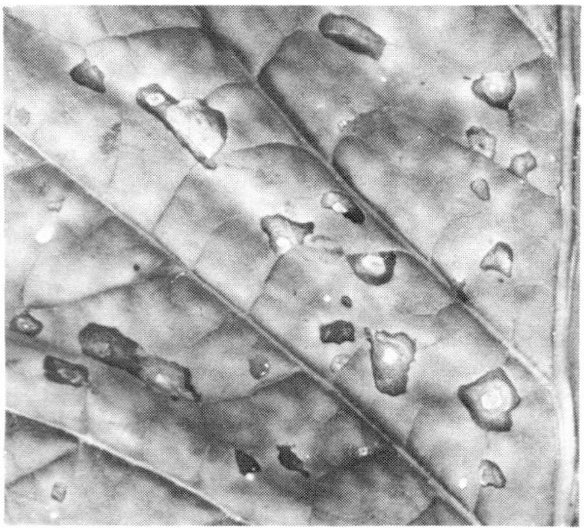

Fig. 258. Typical frog eye.

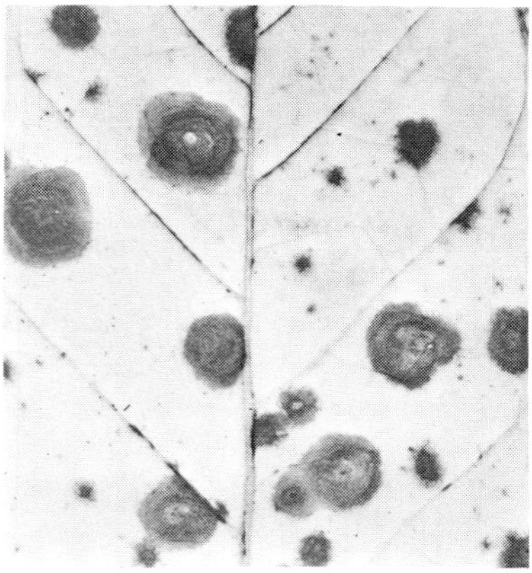

Fig. 259. Severe frog eye on lower leaf of tobacco lacking in nutrients.

Frog eye is common on tobacco wherever it is grown. It is of minor importance but it does cause some deterioration of the leaf during curing. This trouble is known as black barn spot.

Symptoms

Spots on the lower leaves are roughly circular, $\frac{1}{4}$-$\frac{1}{3}$ inch in diameter, with white or pale brown centres, and broad brown margins (Fig. 259). On upper leaves the spots are smaller and more angular (Fig. 258). If infection takes place just before harvest, small green spots develop on the cured leaves. Otherwise the spots are black. Target spots may also develop on the bracts, calyces, and capsules.

Disease Cycle

The frog eye fungus persists in crop refuse, and it may be seedborne. It is spread by *conidia* produced in tufts on stalks *(conidiophores)* on the surface of spots. Infection commences in seedbeds. It is quite possible that spores are in the air the year round.

The disease is worst during hot, humid weather. It may also be favoured by deficiencies of phosphorus and potash, and excessive nitrogen.

Control

1. Check the soil fertility in very severe cases.
2. Do not grow tobacco for more than 2 years on the same land.
3. Spray seedlings with Kopi, Cuprox, or Vitagran (2 oz/3 gallons) every 5 days in the seedbed. Spray the crop with the same fungicide every 10 days during weather that is very favourable for frog eye.

References: 13, 37, 50.

MARGINAL SCORCH

Potash Deficiency

This disorder is fairly common in tobacco planted close to the sea, and also on hill soils of marginal fertility.

Symptoms

The edges of the lower leaves turn yellowish-green and scorch (Fig. 260). Some shothole may be associated with the condition, and diseases such as frogeye may be more severe. Scorched leaves are useless for curing because they do not ripen properly.

Fig. 260. Marginal scorch found at Vatukarasa, Fiji.

Cause

Marginal scorch is generally attributed to a potash deficiency induced by excessive alkalinity of old "reef" soils. It can also occur on well cultivated soils if inadequate fertilizer is supplied to the crop.

Control

1. Have the soil analysed and adjust the fertilizer as indicated.

References: 37, 50.

Tobacco Mosaic Virus

Mosaic is common in all tobacco growing districts. The losses from this disease are not serious among experienced growers.

Symptoms
Leaves show a light and dark green mosaic (Fig. 261). They may also become blistered and twisted. The plants are sometimes stunted and the cured leaves are of low quality.

Fig. 261. Young tobacco leaf showing mottling and twisting caused by a mosaic virus infection.

Disease Cycle
Mosaic is caused by a virus that is transmitted very easily from plant to plant by the sap. The virus can live in dry tobacco for a very long time. If workers smoke or chew tobacco while they are handling seedlings, they will likely infect a very high percentage of the crop. The virus can also be transmitted by animals, machinery, and soil. It can live over from one growing season to the next in solanaceous weeds such as Thorn Apple *(Datura stramonium)*, and nightshade *(Solanum nigrum)*, and in crop refuse and soil.

Control
1. Stop workers from smoking and get them to scrub their hands with strong soap and water with 0.5% Teepol before handling seedlings.
2. Inspect seedbeds and fields and remove suspicious plants.
3. Rotate tobacco with non-solanaceous crops.

References: 37, 50, 74.

Meloidogyne incognita

Fig. 262. Wilted tobacco seedling with heavy infestation of root knot nematodes.

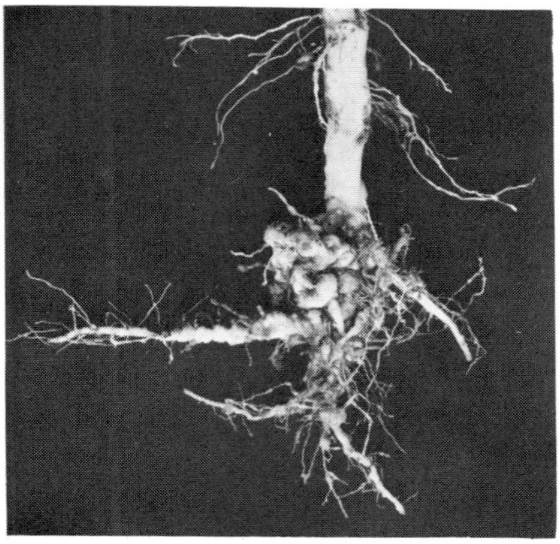

Fig. 263. Swellings on tobacco roots.

Root knot is serious where there is a shortage of land suitable for tobacco and where there is no opportunity for crop rotation.

Symptoms
Plants are stunted, yellowed, and sometimes wilted in patches in fields (Fig. 262). When they are pulled, the roots show the knotty growths (Fig. 263) characteristic of root knot nematode injury.

Disease Cycle
This is similar to that of root knot on banana, ginger, tomato, and other susceptible crops.

Control
1. Fumigate badly infested areas with ethylene dibromide (EDB) emulsion at the rate of 10 gallons per acre on a 12-inch spacing in the rows, before planting.
2. Avoid highly susceptible plants such as banana, tomato, cowpea, etc., in the rotation. Dalo, tapioca, maize, rice, sorghum are all relatively tolerant.

References: 32, 37, 50, 78.

Unknown

Fig. 264. Veinbanding and interveinal white flecks.

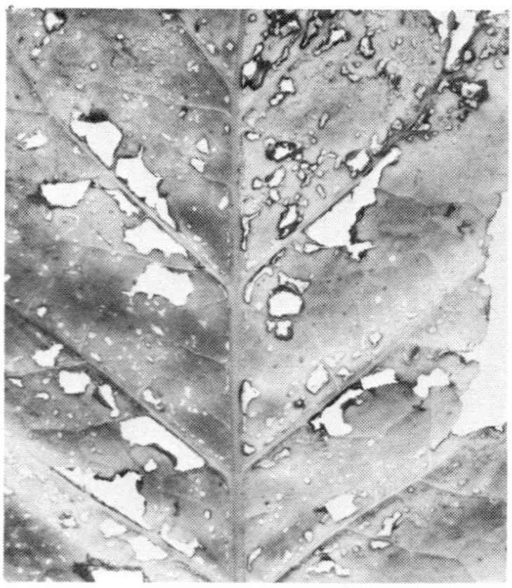

Fig. 265. Veinbanding necrosis and shot-hole.

Several tobacco plants with symptoms of veinbanding were found at Votualevu and Vatukarasa in Fiji in 1966. The economic effect of the disorder was

Symptoms

Tissues along the veins turn white, and those between the veins have numerous white flecks (Fig. 264). On older leaves ready to harvest, the tissues along the veins may die and drop out to produce a shothole effect (Fig. 265). Symptoms are confined to the lower leaves.

Cause

It was possible to transmit veinbanding by rubbing expressed juice from affected leaves on to carborundum-dusted leaves of the variety Samsun. Symptoms appeared in about a week but the plants eventually recovered. The cause is therefore presumed to be viral. The symptoms are similar to those produced by certain strains of tobacco ring spot virus and also by potato virus Y.

Control

1. Rogue infected plants and burn as soon as they are detected.
2. Control weeds in and around tobacco.

References: 11, 37.

ANTHRACNOSE

Glomerella phomoides

Anthracnose, with fruit rots caused by *Alternaria* and *Corynespora,* is the most serious disease of canning tomatoes. It is mainly responsible for high mould counts and off-flavours in the canned product.

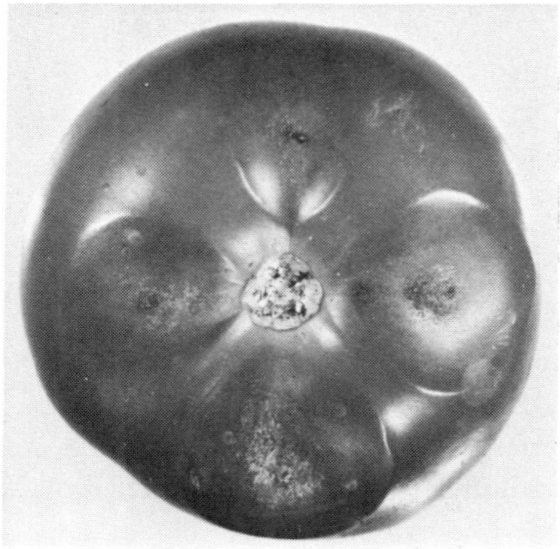

Fig. 266. Tomato anthracnose (produced by inoculation).

Symptoms

Anthracnose commences as small, circular, sunken spots that appear usually on the shoulders of the fruit. They increase in size and run together to cover the fruit surface (Fig. 266). Spots become covered by black pinpoints and a pink slime. Soft rot may follow an attack of anthracnose.

Disease Cycle

The anthracnose fungus may also attack eggplant, chilli, banana and many other ripening fruits. It is also soilborne on decaying organic matter. Infection occurs when the fruit is green and it remains latent or inactive until ripening commences. Infection may also occur through bruises and insect punctures or wounds. *Conidia* produced in great masses in the black pinpoints *(acervuli)* are splashed about by rain from fruit to fruit.

Control

1. Use a 2-year rotation with non-solanaceous crops such as beans, rice, maize, peanuts.
2. Spray with maneb (1 oz/3 gallons) as for control of leaf mould and leaf spots.

References: 14.

BACTERIAL SPOT

Xanthomonas vesicatoria

Bacterial spot is widely distributed anywhere tomatoes are grown. The disease is also common on pepper. It lowers the quality of the fruit.

Fig. 267. Bacterial spot on the variety Red Cloud.

Symptoms

The most obvious symptoms are on the fruit. Spots first appear as small, black, raised pimples surrounded by watersoaked areas. Later the raised portion sinks to form pitted and scab-like lesions (Fig. 267).

Disease Cycle

During rain bacteria ooze from the surface of spots on the fruit and leaves, and they are splashed on to healthy plants. The bacteria enter through stomata or breathing pores, or through wounds. They may be seedborne, or carried in tomato refuse in the soil. Solanaceous weeds such as nightshade *(Solanum nigrum)* may also carry the pathogen. Best weather conditions for disease development are frequent rains and air temperatures between 75° and 86°.

Control

1. Eradicate the bacteria from seed by soaking for 25 minutes in hot water held at 122°F.
2. Buy seed from a source that is certified free of bacterial spot.
3. Spray seedlings in the seedbed every 5 days with a mixture of fixed copper (Kopi or Cuprox) and streptomycin, in water. The proportions of the mixture are 4 lb of fixed copper and 100 ppm of streptomycin. The spray is mixed at the rate of 5 lb in 80 gallons of water and it is applied at 2-3 gallons per 1000 sq ft of bed.
4. Rotate tomatoes with non-solanaceous crops.

References: 14.

Pseudomonas solanacearum

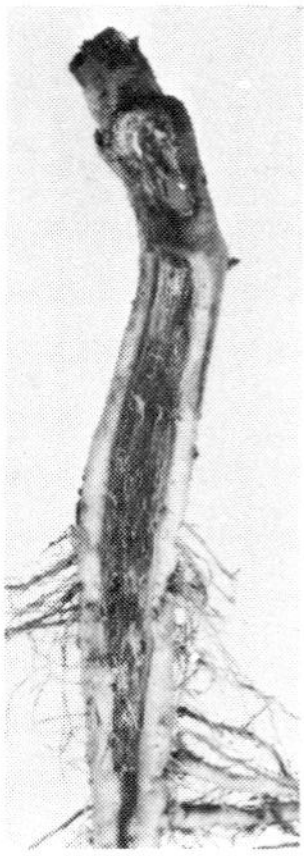

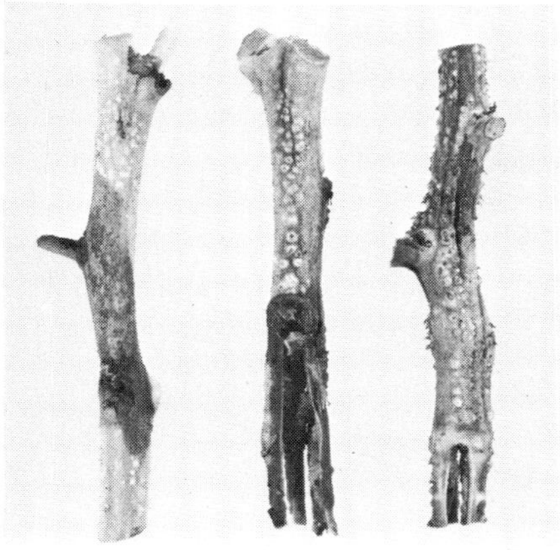

Fig. 268. Wilted tomato plant (above left).

Fig. 269. Vascular browning (above right).

Fig. 270. Adventitious roots on infected stems (left).

Bacterial wilt is a most serious disease of tomatoes in Fiji, Tonga, and Samoa. Tomatoes escape sufficiently during the cool season to produce a crop in Fiji and Tonga.

Symptoms

The lower leaves turn slightly yellow and the plant wilts (Fig. 268). If the plant is pulled, the roots show a browning of the vascular tissue and the pith (Fig. 269). The browning is accompanied by a slimy grey ooze. Abundant adventitious roots (Fig. 270) form on the stems right up to flower height.

Incidence of wilt varies from field to field and from plant to plant. Highly acid soils produce more wilt than alkaline soils.

Disease Cycle

The wilt bacteria cannot enter uninjured roots. They probably get in most infections through the microscopic wounds produced by the feeding punctures of root knot and other nematodes. Any kind of injury may serve as a point of entry. If the soil is wet and the plants are young, they are invaded more rapidly and symptoms will show up within 8-10 days after infection. When infected plants die and decay, the bacteria are liberated into the soil in great numbers.

The wilt organism can attack a wide host range of cultivated plants and weeds, including chillis, tobacco, potatoes, eggplant, peanuts, cowpeas, and many weeds belonging to the tomato family. The most important carrier of the disease in Fiji and probably elsewhere in the South Pacific is the weed Prickly Solanum *(Solanum torvum).*

Best conditions for disease development are a soil pH between 4.5 and 5.5, and soil temperatures between 80° and 85°F.

Control

1. Rotate tomatoes with non-susceptible crops such as rice, maize, sorghum, cabbage, haricot beans, and avoid eggplant, etc.
2. Keep up the soil fertility and apply lime or coral sand as recommended by the Soil Chemist.
3. In small gardens, graft tomatoes on to Prickly Solanum root-stocks if time and local weed control legislation permit.
4. Experiment with resistant varieties available locally.

References: 42, 68.

BLOSSOM-END ROT

Water and Nutrient Imbalance

Blossom-end rot is common where tomatoes are grown after a fallow of weeds or grasses, or where growers have been using high nitrogen fertilizers.

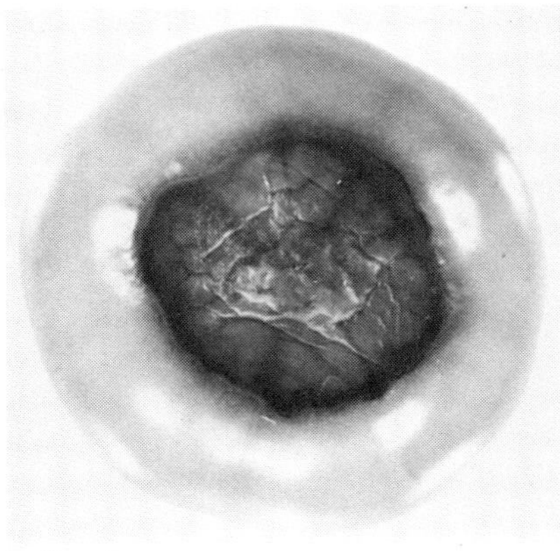

Fig. 271. Blossom-end rot.

Symptoms

A watersoaked area develops at the blossom end of the fruit. This area darkens and collapses (Fig. 271) to form a sunken area covering the entire bottom of the green fruit. It is black, leathery, and it may become invaded by secondary fungi and bacteria. Sometimes the outer tissues are normal but the inside of the fruit has collapsed and turned brown.

Cause

Blossom-end rot is found frequently on highly acid soils that are lacking in calcium, and also where there is a nitrogen/potash imbalance in favour of nitrogen. When nitrogen is applied in excess, blossom-end rot will surely develop. It is common also when heavy rain follows a drought.

Control

1. Check soil fertility and apply a balanced fertilizer.
2. Apply lime if the soil is extremely acid.
3. Keep seedlings growing steadily after transplanting and do not leave them too long in the seedbed.

References: 14, 36.

BUCKEYE ROT

Phytophthora nicotianae var. *parasitica*

Buckeye rot was observed once in American Samoa where it was mistaken for late blight. The pathogen is very common in the South Pacific.

Fig. 272. Buckeye rot with profuse sporulation of *Phytophthora* (produced by inoculation).

Symptoms

Young plants damp off or they develop stem canker. Leaf spots are grey, about ½ inch in diameter and blotchy, like those caused by late blight. The edges of the spots are light green, and slightly watersoaked.

Spots on the fruit consist of pale brown concentric rings. They may be very small or they may involve the whole fruit. Young fruits shrivel and turn black. Large fruits may develop a soft rot. In wet weather, the leaf, stem, and fruit lesions are covered by a white growth of fungous spores (Fig. 272).

Disease Cycle

The pathogen is a water mould also involved in papaw root and fruit rot, dalo corm rot, pineapple heart rot, and damping-off of many vegetables. It is soilborne and it is spread by *zoospores* produced in *sporangia* on the surface of leaf, stem, and fruit lesions. Under adverse weather conditions, e.g., dry weather, *oospores* may be formed. These also germinate to produce sporangia and *zoospores*. They swim about in water droplets and infect fruits and stems through minute injuries. Fruit may be invaded also through the stem end.

Best conditions for infection are 75-86°F. or hot, wet weather.

Control

1. Drench seedbeds with captan (1 oz/3 gallons), or fumigate with formaldehyde (see control of damping-off of all crops).
2. Spray with maneb (1 oz/3 gallons) as for control of other leaf and fruit diseases.

References: 14, 36.

CORYNESPORA LEAF SPOT

Corynespora cassiicola

This is a minor disease of tomatoes found mainly in the Sigatoka Valley of Fiji. It may contribute to defoliation by leaf mould but the effects on the fruit are more serious. The fruit rot reduces quality and the canned product has a mouldy taste.

Fig. 273. Corynespora leaf spot on the variety Red Cloud.

Symptoms

Circular target spots about ¼ inch in diameter, consisting of concentric rings (Fig. 273) of grey alternating with brown develop on the lower leaves. They are often covered by a reddish-brown mould. Spots also develop on the shoulders of the fruit. If sun scald develops following defoliation by leaf mould, fruit rot may occur and cracks in the fruit may also be invaded. Long grey spots also occur on the pedicels and stems of plants.

Disease Cycle

The leaf spot fungus, *Corynespora cassiicola,* has a very wide host range including eggplant, pawpaw, cucurbits, legumes, and it may also be soilborne. The disease on tomato is often confused with early blight, caused by *Alternaria solani*, which attacks potatoes and solanaceous weeds. *Conidia* produced on long, slender, dark-brown conidiophores spread the disease.

The leaf-spotting phase of the disease may be associated with certain soil deficiencies, notably magnesium and potash. The best conditions for disease development are encountered during warm, moist weather.

Control

1. Rotate tomatoes with non-solanaceous crops such as cabbage, rice, maize, and sorghum.
2. Keep solanaceous weeds under control.

3. Spray with mancozeb as for the control of leaf mould.
4. Check soil fertility.

References: 28, 86.

LEAF MOULD

Cladosporium fulvum

Leaf mould is very serious in Fiji and Tonga during the cool season. It greatly reduces yield by defoliating the plants. Fruit is of low quality because it does not ripen properly.

Fig. 274. *Left:* Symptoms of mould on the upper side of a leaflet; *right:* mould growth on the lower side of a leaflet.

Symptoms

The upper sides of the leaves show a yellowish-green mottling and the lower sides are covered with a greenish-purple mould (Fig. 274). As the disease advances, the leaves, starting at the base of the plants and progressing upwards, turn yellow and fall off. Sometimes spots similar to those on the lower sides of the leaves can be found on the young stems and fruit pedicels.

Disease Cycle

Spores of the fungus are produced in great numbers in the greenish-purple layer on the undersides of the leaves. The spores are very light and they can be spread great distances by wind and rain. The fungus can also live in tomato trash from one crop to the next and it can be seedborne. Best conditions for disease development occur during the cool season, with temperatures between 70° and 77°F., and a relative humidity of 95% and above.

Control

1. Spray with mancozeb (Dithane M-45) at a concentration of 1 oz/3 gallons every 10 days, starting when the plants are about 9 inches high and continuing until about 2 weeks from harvest.

References: 14, 36.

MOSAIC

Tobacco Mosaic Virus

Mosaic can be quite serious everywhere tomatoes are grown, because of the high proportion of labourers who smoke tobacco and transmit the disease.

Fig. 275. Mosaic on the variety Red Cloud.

Symptoms

Irregular patches of light green or yellow develop on the leaves. The light areas do not grow as fast as the normal green areas, with the result that the leaves pucker and twist (Fig. 275). If seedlings are infected soon after emergence, they may die. If plants are infected when they are mature, the effect may be very mild. If tobacco and cucumber mosaic infect tomatoes at the same time, black streaks of dead tissue show up on the stems and on the fruit.

Mosaic infection of fruits may be symptomless. If plants are infected early, the fruit is small and deformed, and the carpel walls may show necrotic areas (grey wall).

Disease Cycle

If labourers smoke or chew tobacco, they are very likely to be a source of mosaic infection. The virus is transferred then from plant to plant by the juice. Injured leaf hairs or leaf surfaces may be points of infection. Animals and machinery can also spread the disease. The mosaic virus is very longlived outside tomato and tobacco plants. It can remain alive in dried leaves for years, and it can survive in the soil for more than 4 months on tomato and chilli seeds. Recently it has been demonstrated that tomato mosaic is seed transmitted, but this means is not very important. It can also live over from one growing season to the next in solanaceous weeds, such as Thorne Apple *(Datura stramonium)* and nightshade *(Solanum nigrum)*.

Best conditions for disease development are encountered during cool, cloudy weather. During dry, sunny weather, symptoms may be masked.

Control

1. Insist on a source of seed that has been certified virus free.
2. Inspect seedbeds carefully and remove infected plants.
3. Stop workers from smoking. Get them to scrub their hands with strong soap and water with 0.5% Teepol or some other detergent before handling seedlings.

References: 11, 14, 50, 74.

ROOT KNOT

Meloidogyne incognita

Root knot is serious on tomatoes probably everywhere they are grown in the tropics. It is probably closely linked with bacterial wilt because the feeding points of nematodes provide infection points for bacteria.

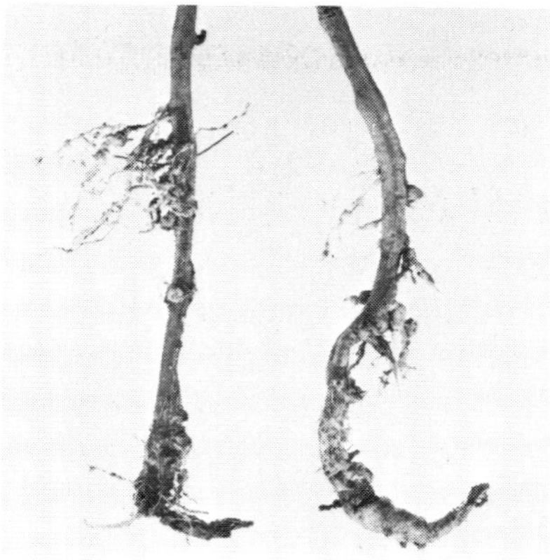

Fig. 276. Severe root knot on the variety Red Cloud.

Symptoms
They are the same as those of root knot in other plants (Fig. 276). During the dry season, wilting is often a very good indication of severe nematode infestation.

Disease Cycle
This is the same as that outlined for root knot of tobacco and carrots.

Control

1. Avoid solanaceous plants and peanuts in the rotation. Use maize, sorghum, and rice as alternate crops.
2. Control weeds.
3. Try soil fumigation with EDB (ethylene dibromide) at the rate of 10 gallons per acre. Inject well prepared soil at a 6-inch depth on a 12-inch spacing.
4. Try the root knot-resistant varieties available locally.

References: 18, 19, 36, 78.

SHOESTRING

Cucumber Mosaic Virus

Shoestring has been observed on rare occasions in Suva backyard gardens. It may be quite common elsewhere.

Symptoms
Leaves are quite narrow and in severe cases they resemble shoestrings (Fig. 277). Leaves are also curled, and light green in colour. Fruits are smaller than normal. Setting of fruit on the upper trusses may be prevented.

Fig. 277. Symptoms of shoestring on tomato leaf.

Disease Cycle
Shoestring is caused by the cucumber mosaic virus. If both this virus and tobacco mosaic virus infect tomato plants simultaneously, extreme shoestring, resembling dodder, may develop. The source of virus is probably infected weeds in seedbeds or plantations. It is easily transmitted by handling first infected plants, and then tomatoes. The virus may also be carried by aphids. Cucumber mosaic has many hosts.

Control
1. Make sure that the tomato seedbed is free of weeds and aphids.
2. Remove suspicious plants from the seedbed and wash the hands thoroughly with strong soap and water before handling healthy plants.
3. Do not grow susceptible crops, e.g., cucurbits, passion fruit, cowpeas, close to tomatoes.
4. Control weeds in tomato fields.

References: 11, 14, 74.

STREAK

Mixed Viruses

Fig. 278. Bushy, stunted plant with very small leaves.

Fig. 279. Streaks of dead tissue on tomato stem and petiole.

Streak was found in a tomato field in the Sigatoka Valley in 1966, and in the vegetable gardens of the Fiji School of Agriculture at Koronivia in 1967. Incidence of the disease was very low.

Symptoms

Plants are stunted and they have a bushy appearance (Fig. 278). The upper leaves curl downwards and inwards. Leaves are very small, sometimes resembling fern leaves, and they are yellowish. The veins are translucent at first and then they turn black. Necrotic streaks can be seen on petioles and stems (Fig. 279). The top of the plant dies back and the upper trusses fail to set fruit. The lower trusses set undersized, deformed fruit.

Disease Cycle

The viruses involved in streak have not been determined. The local disease is suspected to be due to a mixed infection of cucumber mosaic and tobacco mosaic, both of which have been found separately in tomatoes. Both of these viruses are highly infectious and they can be transmitted by handling of plants during cultivation or tying to stakes.

Control

1. Pull and destroy diseased plants as soon as they are detected.
2. Apply the same precautions as for control of other virus diseases.

References: 11, 14, 74.

Defoliation

Sunscald is not, strictly speaking, a distinct and separate disease of tomatoes. It is a condition that follows defoliation caused by fungus diseases.

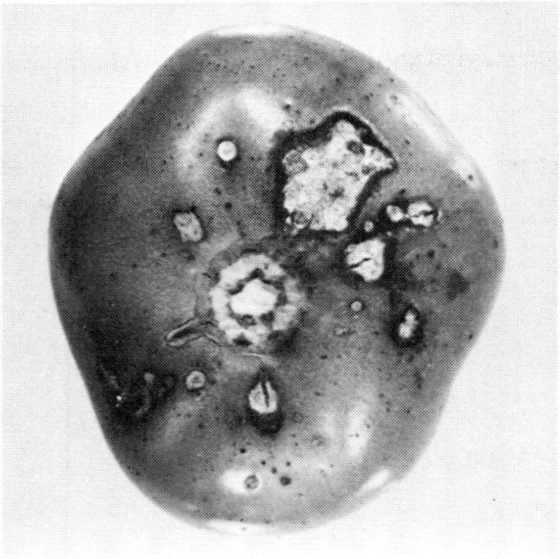

Fig. 280. Early stage of sunscald on the variety Red Cloud.

Symptoms
The shoulders of the green fruit develop sunken grey or white patches encircling the pedicel (Fig. 280). These areas often turn reddish-brown or black following invasion by secondary fungi. The fruit may rot completely.

Cause
Defoliation by leaf mould *(Cladosporium fulvum)*, early blight *(Alternaria solani)*, Corynespora blight *(Corynespora cassiicola)*, or leaf spot *(Septoria lycopersici)*, can all cause sunscald. It is more common during the dry season, when cloud cover is less and foliar diseases are more prevalent.

Control
1. Control foliar diseases by application of fungicide.

References: 14.

YELLOWS

Magnesium deficiency

Tomato yellows is quite common on sandy, acid soils low in fertility. Affected plants appear to be more susceptible to leaf-spotting diseases.

Fig. 281. *Left:* Healthy leaflet; *right:* leaflet with yellows.

Symptoms
Symptoms begin to show on the lower leaves and then they progress upwards. The interveinal tissues turn yellow, but the veins remain banded by green (Fig. 281). Leaves tend to curl upwards.

Cause
Soil samples taken from affected areas have indicated a deficiency of magnesium. The soil upon which yellows was found is known to be fairly acid and low in fertility. Similar symptoms were evident on haricot bean and lettuce.

Control
1. Check soil fertility.
2. Apply fertilizer and lime in accordance with the recommendations of the Soil Chemist.

References: 14, 84.

LEAF SPOT

Cercospora sp.

Fig. 282. Early stage of yam leaf spot.

Fig. 283. Advanced stage of yam leaf spot.

This leaf-spotting disease is common on yam in Fiji. It does not appear to do much damage.

Symptoms
Leaf spots begin as small brown or yellowish areas that enlarge to become angular black spots about ¼ inch across (Fig. 282). These may cover the whole leaf and their centres may fall out (Fig. 283). Elongated sunken spots may also appear on the petioles and the vines.

Disease Cycle
Conidia produced on both sides of the leaves spread the disease during rainy weather. New spots develop within 6-12 days. The disease appears to be more common during the wet warm season.

Control
1. Spray with maneb or captan (1 oz/3 gallons) every 10 days where the disease is severe.

References: 13, 14.

WILT AND STEM ROT

Unknown

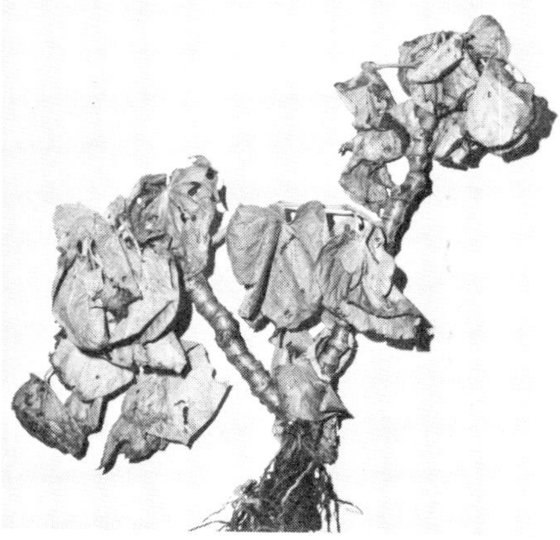

Fig. 284. Wilted plant of a green-stemmed variety of yaqona.

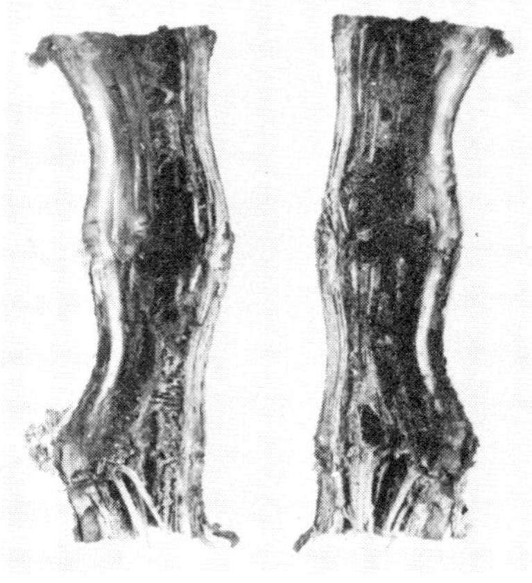

Fig. 285. Internal rot of yaqona stem.

Wilt and stem rot was first observed in the Suva Peninsula of Fiji in 1936. It is not quite common in the wet zones of Viti Levu, Vanua Levu, Taveuni, and Kadavu. It seriously limits production.

Symptoms
Leaves of mature plants wilt (Fig. 284) and a black rot appears on the lower parts of the stems. If the stems are cut, they show an internal rot (Fig. 285), and a blackening along the vascular system. The roots are also rotted.

Cause
The exact cause is unknown. Bacteria have sometimes been isolated from the discoloured vascular system, but they have never been demonstrated to cause disease. Recently a species of water mould, *Pythium splendens,* was isolated from rotted roots, but it failed to cause rotting of inoculated cuttings.

The disease is worst in poorly drained soils in the wet zones. It is probably confused with the effects of malnutrition where yaqona is planted on red hill soils.

The green-stemmed varieties are more susceptible than those with purple pigment.

Control
1. Plant yaqona in well-drained black soil.

References: 61.

Crop	Fungicide	Rate of Application	
Forage Crops			
Grasses	Captan	6 oz/100 lb	2 tsp/lb
Legumes	Spergon (Coversan)	10 oz/100 lb	3 tsp/lb
Field Crops			
Maize	Captan, thiram	2 oz/100 lb	½ tsp/lb
Rice	Ceresan	2 oz/100 lb	½ tsp/lb
Sorghum	Ceresan	2 oz/100 lb	½ tsp/lb
Peanut	Captan	4 oz/100 lb	1 tsp/lb
Tobacco	Silver nitrate	Soak for 5 min in .001	
Vegetable Crops			
Bean	Spergon (Coversan)	4 oz/100 lb	1 tsp/lb
*Cabbage	Captan, thiram	3 oz/100 lb	1 tsp/lb
Carrot	" "	5 oz/100 lb	1½ tsp/lb
*Cauliflower	" "	3 oz/100 lb	1 tsp/lb
Chinese cabbage	" "	3 oz/100 lb	1 tsp/lb
Cucurbits	" "	3 oz/100 lb	1 tsp/lb
Eggplant	" "	4 oz/100 lb	1 tsp/lb
Lettuce	" "	3 oz/100 lb	1 tsp/lb
Onion	" "	4 oz/100 lb	1 tsp/lb
Pepper (chilli)	" "	4 oz/100 lb	1 tsp/lb
Pea	Spergon (Coversan)	3 oz/100 lb	1 tsp/lb
Potato	5% thiram Dust	8 oz/100 lb	
Pulses	Spergon (Coversan)	4 oz/100 lb	1 tsp/lb
Tomato	Captan, thiram	5 oz/100 lb	1½ tsp/lb
Silver beet	" "	10 oz/100 lb	3 tsp/lb
Tree Crops			
Avocado	Captan, thiram	4 oz/100 lb	1 tsp/lb
Citrus	" "	3 oz/100 lb	1 tsp/lb
Coffee	" "	3 oz/100 lb	1 tsp/lb
Cocoa	" "	4 oz/100 lb	1 tsp/lb
Papaw	" "	5 oz/100 lb	1½ tsp/lb
Pepper (black)	" "	5 oz/100 lb	1½ tsp/lb
Passion Fruit	" "	4 oz/100 lb	1 tsp/lb
Tea	" "	3 oz/100 lb	1 tsp/lb

* Cabbage, Cauliflower: for control of black rot, soak seed in warm water for 10 minutes at 110°F., dry in the sun, and then apply seed protectant.

References: 3, 52, 53, 69.

SEED TREATMENT

All crop seeds should be treated with fungicide as a precaution against seed rot, damping off, seedling blight, and crown rot or canker in the post-emergence phase of growth. Captan and Thiram dusts are the most popular chemicals used for this purpose. Spergon is best for treatment of legumes, because it has little effect on nitrogen-fixing nodule bacteria that are often added to seed at the time of sowing. For cereal crops the dry mercurial seed dressings are best because they can penetrate the glumes and kill fungi on the seed surface.

Small quantities may be treated by sprinkling the fungicidal dust on to the seed and shaking it vigorously in a large plastic bag until each seed is well coated. Large quantities should be treated by shovelling to and fro on a flat concrete platform or floor. A protective mask should be worn to avoid breathing irritating chemical dust.

Seed stored for any length of time after treatment and before sowing must be kept out of reach of children and animals.

SOIL FUMIGATION

Soil fumigants in the form of a gas or a water-soluble material are used for control of damping-off in seedbeds, nurseries, and vegetable plots. A special class, the nematocides, are used on *high-value annual crops,* e.g., tobacco, ginger. The economics of the use of nematocides for control of nematodes in bananas must be worked out.

Recommendations

1. Put soil into good seedbed condition down to plough or spade depth. Fumigants will not penetrate large lumps or unrotted roots.
2. Apply treatment when soil is in good condition for seed germination, or when it is at 50% or its water-holding capacity.
3. Apply more fumigant to clay and organic soils than to sandy soils.
4. Fumigate preferably when the soil temperature is 60-75°F.
5. Concentrate treatment where troubles are worst, e.g., root knot nematode infestation of tobacco, instead of trying to treat the whole field.
6. Exercise strict supervision of the use of methyl bromide (Dowfume MC-2).

References: 36, 52, 53, 69.

TRADE NAME	CHEMICAL NAME and CONCENTRATION*	RATE and METHOD of APPLICATION
DOWFUME MC-2	Methyl bromide + Chloropicrin marker	1-2 lb/100 sq ft applied under a plastic sheet; topdress soil with high-N fertilizer before sowing.
SHELL SOIL FUMIGANT NEMAFUME 15	Ethylene dibromide (EDB) 15% conc.	3 pt/1000 sq ft or 15 gal/acre by injection before planting.
FORMALDEHYDE	Formalin 40% conc.	2 pt/sq ft of 1 gal/20 gal dilution; cover with tarpaulin or bags for 24 hr.
FUMAZONE (NEMAGON)	Dibromodichloropropane (DBCP) 75% or 85% w/v active ingredient.	1-3 gal/acre by injection before planting; use 10% (1/10) dilution; dip banana corms in 40 oz/100 gal for 1 minute.
TERRACHLOR (PURASOIL, QUINTOZENE)	Pentachloronitrobenzene (PCNB) 75% conc.	1 gal/5 sq yd of 1 lb/10 gal dilution applied as drench. 1 lb dust/100 ft of row, treatment before planting: work in 2-4" deep.
STEAM	–	*Sterilization:* pass live steam into soil until held at 180°F for ½ hour. *Pasteurization:* pass low-pressure steam into soil until all parts reach 140-160°F for ½ hour.

* Concentrations given are those now available in Fiji. In the future formulations with other concentrations may be imported. Be sure to read the label and follow manufacturers' directions.

TREATMENT/PLANTING INTERVAL	DISEASES CONTROLLED	SENSITIVE CROPS
2-7 days, depending on moisture content	Damping-off, nematodes, insects, and some weed seeds in seedbeds.	None
2-7 days, depending on moisture content	Root knot nematodes in the field.	Onion
10 days	Damping-off, root and collar rot.	None
10 days for each gallon of concentrate.	Burrowing and other parasitic nematodes in the field.	Kumala, Lettuce, Onion, Potato, Silver Beet, Tobacco
None	Damping-off and collar rot caused by Rhizoctonia & Sclerotium fungi in seedbeds and nurseries; stem rot of peanuts, and vegetables in the field.	Cucurbits, Tomatoes
48 hours	Damping-off, nematodes insects, and weed seeds in seedbeds.	None

Measure of Capacity *(Fluid)*

20 fluid ounces	= 1 pint
2 pints	= 1 quart
4 quarts	= 1 gallon
1 fluid ounce	= 2 tablespoons
1 tablespoon	= 4 teaspoons
1 teaspoon	= 70 drops (eyedropper)
1 gallon (U.S.)	= ⅚ gallon (Imperial)
1.2 " "	= 1 " "

Conversions

Fluids

1 fl oz	= 28.4 c.c.
1 pint	= 568.25 c.c.
1 gallon	= 4546 c.c.
35.2 fl oz	= 1000 c.c.
1 litre	= 1.76 pints

Powders

1 lb	= 453.6 grams
1 oz	= 28.4 grams
35.27 oz	= 1 kilogram
2.2 lb	= 1 kilogram

Measure of Small Quantities of Wettable Powders

Captan	9 rounded teaspoons	= 1 oz
Cuprox	5½ " "	= 1 oz
Karathane	6 " "	= 1 oz
Kopi	5½ " "	= 1 oz
Maneb	11 " "	= 1 oz
Perenox	5 " "	= 1 oz
Cupoxide	5 " "	= 1 oz
Copper Sandoz	5 " "	= 1 oz
Vitagran	5½ " "	= 1 oz
Zineb	11 " "	= 1 oz

Capacity of Knapsack Sprayers and Gallonage per Acre

1 Rega or other large knapsack sprayer	= 3 gallons of water
1 Standard small knapsack sprayer	= 2 gallons of water
14 large sprayers	= 40 gallons per acre
20 small sprayers	= 40 gallons per acre
27 large sprayers	= 80 gallons per acre
40 small sprayers	= 80 gallons per acre

GLOSSARY

Acervulus — A flat, open bed of generally short conidiophores with conidia at the tips.

Anthracnose — A plant disease having characteristic limited lesions, usually caused by species of *Glomerella* and their Colletotrichum or Gloeosporium states.

Ascus — A sac-like cell, containing ascospores.

Ascospore — A spore produced inside an ascus.

Bacterium — A microscopic, one-celled or filamentous organism, smaller than a fungus and larger than a virus.

Basidium — A club-shaped spore-bearing structure, usually bearing 4 basidiospores.

Basidiospore — A spore produced on a basidium.

Blight — A rapid discoloration of tissue over certain portions of the plant.

Bracket fungus — A fungus growth that resembles a flat, circular bracket, usually associated with heart, trunk, or root rots of trees.

Canker — A dead area on a trunk or stem, surrounded by living tissue; slow rots of the outer parts of stems and sometimes of fruits.

Chlamydospore — A thick-walled spore formed by the rounding-off of certain sections of a hypha; usually resistant to unfavourable conditions.

Chlorosis — A subnormal development of chlorophyll resulting in a light green, yellow, or white condition.

Collar rot — A rot at the base of woody stems.

Conidiophore — A simple or branched hypha, on which conidia are produced.

Conidium — A spore that is asexually produced and which germinates by a germ tube.

Damping-off — Death of seedlings associated usually with rapid drying; death of tissue at the ground line, followed by toppling over.

Dieback — Progressive dying of tissue from the tip to the base of stems, usually of trees.

Dodder — A parasitic flowering plant.

Enation — A small outgrowth from a leaf or stem.

Emulsifier — A chemical used to disperse oil in water, or water in oil.

Emulsion — A mixture of oil and water, usually used as a base for addition of a fungicide or insecticide, or another chemical.

Eye spot — A plant disease characterized by grey or white spots with black or dark brown margins.

Fumigant — A chemical used in the form of a gas or a water-soluble material for the killing of soilborne pathogens in seedbeds, vegetable plots, bananas, and *high value* annual crops.

Fungicide — A chemical that is used to kill fungi.

Fungus	A microscopic plant lacking chlorophyll, parasitic or living on dead matter.
Fasciation	A plant disease which is manifested by plant organs being massed or joined side by side.
Gall	An overgrowth induced by an increase in size or number of cells as a result of abnormal stimulation.
Germ tube	The hypha produced by a germinating spore.
Gummosis	A plant disease having secretion of gum as a well-marked symptom.
Hairy root	Development of a large number of small roots in a restricted area on a root.
Halo	A narrow, chlorotic zone around a spot, due to interference with chlorophyll formation.
Haustorium	A special hyphal branch, especially one within a cell of the host, for absorption of food; also applied to the holdfasts and absorbing organs of parasitic flowering plants.
Host	The plant on, or in which, a parasite is living and obtaining its food.
Hydathode	An open pore located on the leaf margin at the end of a vein, through which water droplets may exude.
Infection	The establishment of a disease-producing organism on or within a host.
Knot	A warty overgrowth, usually on a root.
Lesion (spot)	A localized injury caused by an infection. It is similar to a canker, but much smaller.
Melanose	A plant disease characterized by the presence of black spots, fruiting bodies, etc.
Mildew	A plant disease in which the pathogen is seen as a growth on the surface of the host.
Mosaic	A disarrangement of the chlorophyll content of plant tissues, especially leaves, causing dark green and light green areas to form a mosaic-like pattern. Some distortion may be involved.
Mottle	Occurrence of a grouping of light-green or yellowish-green areas, usually round, without distortion.
Mycelium	A mass of hyphae.
Nematode	A microscopic worm.
Oospore	A resting spore from a fertilized female organ of the water moulds.
Parasite	An organism living on or in, and getting its food from another living organism.
Pathogen	A parasite able to cause disease in a particular plant or plants.
Perithecium	A closed flask-like fruiting body with a pore *(ostiole)* through which ascospores are ejected.
Psorosis	A plant disease in which there is much scaling of bark.

Pustule	A local elevation of the plant epidermis, which may rupture to expose the causal agent.
Ring spot	Yellowish rings on the leaves with green tissues inside the ring. This is usually associated with virus diseases.
Rosette	A rose-like cluster of organs, due to shortening of stems and simultaneous development of a number of buds.
Russetting	Brownish, roughened areas on the surface of fruits, caused by disease, pests, or sprays.
Rust	A plant disease with "rusty" symptoms, due to the presence of a large number of pustules of brown fungal spores.
Scab	A crust-like lesion.
Sclerotium	A firm, often rounded mass of hyphae, with or without the addition of host tissues or soil, normally having no spores in or on it; usually resistant to unfavourable conditions.
Shothole	A leaf-spotting disease characterized by holes made by the dead parts falling out.
Sporangiophore	A simple or branched hypha, upon which sporangia are produced.
Sporangium	An organ producing spores inside it by asexual means.
Sporodochium	A cushion-shaped fruiting body that develops under the host surface and pushes up through it to release its conidia.
Spore	The reproductive unit or "seed" of a fungus.
Smut	A disease caused by one of the smut fungi, of which the spores are characteristically powdery and usually black.
Thread blight	A plant disease characterized by the presence of visible thread-like fungus growth.
Toxin	A chemical toxic or poisonous to tissues.
Uredospore	Spore of a rust fungus.
Uredium	The repeating stage of a rust fungus; a rust pustule produced by pushing up through the host epidermis.
Vector	An agent, e.g., insect, that transmits disease-producing organisms.
Vein clearing	A chlorosis along the veins, indicating invasion by a virus.
Virus	A sub-microscopic particle, capable of reproducing itself only in the cells of a living organism.
Wetting agent	A chemical applied to a plant surface which eliminates air thus ensuring that a fungicide or insecticide makes proper contact and gives adequate protection.
Zoospore	A swimming spore produced inside a sporangium.

References: 1, 2, 7, 10, 11, 26, 36, 53, 69, 74, 78.

SELECTED BIBLIOGRAPHY

1. Ainsworth, G.C. Ainsworth and Bisby's Dictionary of the Fungi. 5th Ed. Commonwealth Mycological Institute, Kew, 1963.
2. Alexopoulos, C.J. Introductory mycology. 2nd Ed. John Wiley & Sons, New York. 1962.
3. American Phytopathological Society. Fungicide–Nematocide tests. vol. 22. 1966.
4. Atkins, J.G. Rice diseases. USDA Farmers' Bull. No. 2120. 1958.
5. Aycock, R. Stem rot and other diseases caused by *Sclerotium rolfsii.* North Carolina Agr. Exp. Sta. Tech. Bull. No. 174. 1966.
6. Brien, R.M. *et al.* Diseases and pests of peas and beans in New Zealand and their control. N.Z. Dept. Sci. & Industry Res. Bull. No. 114. 1955.
7. Brooks, F.T. Plant diseases. 2nd Ed. Oxford University Press, London. 1953.
8. Bryan, O.C. Malnutrition symptoms of citrus. Fla. Dept. Agr. Bull. No. 93. 1961.
9. Calpouzos, L. Studies on the Sigatoka disease of bananas and its fungus pathogen. Atkins Garden & Research Lab., Cienfuegos, Cuba. 1955.
10. Carter, W. Insects in relation to plant disease. Interscience Publishers, New York. 1962.
11. Chamberlain, E.E. Plant virus diseases in New Zealand. N.Z. Dept. Sci. & Industry Res. Bull. No. 108. 1954.
12. Child, R. Coconuts. Longmans, London. 1964.
13. Chupp, C. A monograph of the genus *Cercospora.* Ithaca, New York. 1954.
14. Chupp, C. and A.F. Sherf. Vegetable diseases and their control. Ronald Press, New York. 1960.
15. Cobley, L.S. An introduction to the botany of tropical crops. Longmans, London. 1956.
16. Coffee Research Services. An atlas of coffee pests and diseases. Coffee Board of Kenya, Nairobi. 1962.
17. Colbran, R.C. Effects of treatments of banana "bits" for nematode control on emergence and yield. Qld. J. Agr. Sci. 21: 237-238. 1964.
18. ——. Rootknot nematodes. Qld. Div. Pl. Industry Adv. Leaflet No. 848. 1966.
19. ——. Studies on rootknot nematode control in ginger. Qld. Div. Pl. Industry Bull. No. 218. 1962.
20. Collins, J.L. The pineapple. Leonard Hill, London, 1960.
21. Da Costa, E.W.B. Diseases of papaw. Qld. Dept. Agr. & Stock Pamphlet No. 41. 1944.
22. Del Rosario, Maria Salome E. Diseases of coconut. (Cylostyled brochure prepared for East–West Center, University of Hawaii field training course in plant pathology, Western Samoa, May, 1965).
23. Del Rosario, Maria Salome E., D.A. Benigno, and L.P. Libed. Virus diseases of weeds in the Philippines. I. *Passiflora foetida* Linn. Phil. Agriculturist 58: 95-111. 1964.

SELECTED BIBLIOGRAPHY

23A Del Rosario, Maria Salome E., T. T. Reyes, and A. R. Salgado. Mosaic Virus on the pechay in the Philippines. Philippine Agriculturalist 50: 213-227. 1966.

24. Dickson, J.G. Diseases of field crops. 2nd Ed. McGraw-Hill, New York. 1956.

25. Dirven, J.G.P. and H.A. van Hoof. A destructive virus disease of pangola grass. Tijdschrift für Pflantenziekten 66: 344-349. 1960.

26. Dowson, W.J. Manual of bacterial plant diseases. A. & C. Black, London. 1949.

27. Eden, T. Tea. 2nd Ed. Longmans, London. 1965.

28. Ellis, M.B. Some species of *Corynespora.* CMI Mycological Paper No. 65. 1957.

29. Fawcett, H.S. Citrus diseases and their control. 2nd Ed. McGraw-Hill, New York. 1936.

30. Fiji Dept. of Agriculture Annual Report for 1960. Legislative Council Paper No. 25. 1961.

31. Fiji Dept. of Agriculture Annual Report for 1963. Legislative Council Paper No. 17. 1964.

32. Franklin, M.T. and D.J. Hooper. Plants recorded as resistant to root knot nematodes *(Meloidogyne* spp.*).* Commonwealth Bureau of Helminthology Tech. Comm. No. 31. 1959.

33. Goodey, J.B., M.T. Franklin, and D.J. Hooper. T. Goodey's The nematode parasites of plants catalogued under their hosts. 3rd Ed. Commonwealth Agricultural Bureaux, Farnham Royal, Bucks, England. 1965.

34. Harwood, L.W. Diseases of cocoa. Fiji Agr. Journal 29: 96-98. 1959.

35. Hine, R.B., O.V.Holtzmann, and R.D. Raabe. Diseases of papaya. Hawaii Agr. Exp. Sta. Bull. No. 136. 1964.

36. Holtzmann, O.V. and E.E. Trujillo. (Editors) Tropical plant disease manual. (Cyclostyled brochure prepared for East–West Centre, University of Hawaii field training course in plant pathology, Western Samoa, May, 1965).

37. Hopkins, J.C.F. Tobacco diseases, with special reference to Africa. Commonwealth Mycological Institute, Kew. 1956.

38. Hughes, C.G., E.V. Abbott, and C.A. Wismer. Sugar cane diseases of the world. vol. 2. Elsevier, Amsterdam. 1964.

39. Jacob, A. and H. von Uexküll. Fertilizer use: nutrition and manuring of tropical crops. 3rd Ed. Verlags. für Ackerbau, Hanover. 1963.

40. International Rice Research Institute. The rice blast disease. Proceedings of a symposium at the IRRI. Johns Hopkins, Baltimore. 1963.

41. Ishii, M. and I. Pascual. A virus disease of passion fruit. Hawaii Farm Science 13, No. 4: 9-11. 1964.

42. Kelman, A. The bacterial wilt caused by *Pseudomonas solanacearum.* North Carolina Agr. Exp. Sta. Bull. No. 99. 1953.

43. Klein, H.H. Control of Cercospora leaf spot of bananas with applications of oil sprays based on the disease cycle. Phytopathology 50: 488-490. 1960.

44. Klemmer, H.W. and R.Y. Nakano. Distribution and pathogenicity of *Phytophthora* and *Pythium* in pineapple soils of Hawaii. Plant Disease Reporter 48: 848-852. 1964.
45. Klotz, L.J. Color handbook of citrus diseases. 3rd Ed. University of California Div. of Agricultural Sciences. 1961.
46. Knorr, L.C. Serious diseases of citrus foreign to Florida. Fla. Dept. Agr. Bull. No. 5. 1965.
47. Knorr, L.C., L.F. Suit, and E.P.Ducharme. Handbook of citrus diseases in Florida. University of Florida Agr. Exp. Sta. Bull. No. 587. 1957.
48. Leach, R. Banana leaf spot *(Mycosphaerella musicola)* on the Gros Michel variety in Jamaica. Investigations on the aetiology of the disease and the principles of control by spraying. Govt. Printer, Jamaica. 1946.
49. Leach, R. Report on investigations into the cause and control of the new banana disease in Fiji, black leaf streak. Government of Fiji Legislative Council Paper No. 38. 1964.
50. Lucas, G.B. Diseases of tobacco. 2nd Ed. Scarecrow Press, New York. 1965.
51. Luh, T.H., C. Huang, and H.W. Ream. Grassland potential and current development in Taiwan. Chinese–American Joint Commission on Rural Reconstruction, Animal Industry Series No. 6. 1961.
52. Martin, H. (Editor) Insecticide and fungicide handbook. Blackwell's, Oxford. 1963.
53. Martin, H. The scientific principles of plant protection. 5th Ed. Edward Arnold, London. 1964.
54. McKnight, T. The woodiness virus of the passion vine. Qld. Dept. Agr. & Stock, Div. Pl. Industry Bull. No. 67. 1953.
55. Martin, J.P., E.V. Abbott, and C.G. Hughes. Sugar cane diseases of the world. vol. 1. Elsevier, Amsterdam. 1961.
56. Morwood, R.B. A preliminary list of plant diseases in Fiji. Fiji Agr. Journal 27: 51-54. 1956.
57. ——. Citrus canker. Fiji Agr. Journal 25: 15-16. 1954.
58. Mune, T.L. and J.W. Parham. The declared noxious weeds of Fiji and their control. 3rd Ed. Fiji Dept. Agr. Bull. No. 48, 1967.
59. Oxenham, B.L. Notes on two pineapple diseases in Queensland. Qsld. Dept. Agr. & Stock, Div. Pl. Industry Bull. No. 74. 1953.
60. Padwick, G.W. Manual of rice diseases. Commonwealth Mycological Institute, Kew. 1950.
61. Parham, B.E.V. Wilt disease of yaqona. Fiji Agr. Journal 8: 2-8. 1935.
62. Parham J.W. Plants of the Fiji Islands. Government Press, Suva. 1964.
63. Price, W.C. (Editor) Proceedings of the 3rd conference of the International Organization of Citrus Virologists, São Paulo, Brazil. University of Florida Press, Gainesville. 1965.
64. Purss, G.S. Peanut diseases in Queensland. Qld. Dept. Agr. & Stock, Div. Pl. Industry Adv. Leaflet No. 676. 1962.
65. Queensland Dept. of Agriculture and Stock. The Queensland Agricultural and Pastoral Handbook. vol. 3. Insect pests and diseases. Government Printer, Brisbane. 1951.

66. Raabe, R.D. and O.V. Holtzmann. Studies on the control of papaya anthracnose. Hawaii Farm Science 13, No. 4: 1-2. 1964.
67. Ramakrishnan, T.S. Diseases of millets. Indian Council of Agricultural Research, New Delhi. 1963.
68. Rangaswami, G. Bacterial plant diseases in India. Asia Publishing House, London. 1962.
69. Sharvelle, E.G. The nature and uses of modern fungicides. Burgess Publishing Co., Minneapolis. 1960.
70. Shaw, D.E. Plant pathogens and other micro-organisms in Papua and New Guinea. Dept. Agr., Stock & Fisheries Res. Bull. No. 1. 1963.
71. ——. Diseases of coconut in Papua and New Guinea. Papua and New Guinea Agr. Journal 17: 67-71. 1965.
72. Shorrocks, V.M. Mineral deficiencies in *Hevea* and associated cover plants. Rubber Research Institute, Kuala Lumpur. 1964.
73. Sills, V.E. Notes on the treatment, packing, and cool storage of bananas for export. Fiji Dept. Agr. Sci. Bull. No. 44. 1965.
74. Smith, K.M. A textbook of plant virus diseases. 2nd Ed. J. & A. Churchill, London. 1957.
75. Stover, R.H. Fusarial wilt (Panama disease) of bananas and other Musa species. CMI Phytopathological Paper No. 4. 1962.
76. ——. Leaf spot of bananas caused by *Mycosphaerella musicola:* factors influencing production of fructifications and ascospores. Phytopathology 54: 1320-1326. 1964.
77. Tarr, S.A.J. Diseases of sorghum, sudan grass, and broom corn. Commonwealth Mycological Institute, Kew. 1962.
78. Thorne, G. Principles of nematology. McGraw-Hill, New York. 1961.
79. Trujillo, E.E. Diseases of ginger *(Zingiber officinale)* in Hawaii. Hawaii Agr. Exp. Sta. Circular No. 62. 1964.
80. ——. and R.B. Hine. Papaya root rot and the ' replant problem". Hawaii Farm Science 13, No. 4: 3-5. 1964.
81. Urquhart, D.H. Cocoa. 2nd Ed. Longmans, London. 1961.
82. Wallace, J.M. (Editor) Citrus virus diseases. Proceedings of a conference on citrus virus diseases, Riverside, California. University of California, Div. of Agricultural Sciences, Berkeley. 1959.
83. ——. and R.J. Drake. Induction of woody galls by wounding of citrus infected with vein-enation virus. Plant Disease Reporter 45: 682-686. 1961.
84. Wallace, T. Diagnosis of mineral deficiencies in plants by visual symptoms. H.M. Stationery Office, London. 1961.
85. Wardlaw, C.W. Banana diseases. Longmans, London. 1961.
86. Wei, C.T. Notes on *Corynespora.* CMI Mycological Paper No. 34, 1950.
87. Wellman, F.L. Coffee: botany, cultivation, and utilization. Leonard Hill, London. 1961.
88. Whitaker, T.W. and G.N. Davis. Cucurbits. Leonard Hill, London. 1962.
89. Zaiger, D. and G.A. Zentmyer. Phytophthora canker of cacao in the Caroline Islands. Plant Disease Reporter 49: 565-567. 1965.

90. Zentmyer, G.A. *et al.* Avocado diseases. Calif. Agr. Exp. Sta. Ext. Service Circular No. 534. 1965.

APPENDIX

DISTRIBUTION RECORDS for the SOUTH PACIFIC

BSI – British Solomon Islands
FP – French Polynesia
NC – New Caledonia
NH – New Hebrides
PNG – Territory of Papua-New Guinea
USTT– U.S. Trust Territory of the Pacific

Host	*Disease*	*Pathogen*	*Distribution*
All crops	Stem rot	*Sclerotium rolfsii*	Probably universal; BSI, Fiji, FP, NC NH, PNG, Tonga.
Avocado	Anthracnose and ripe fruit rot	*Glomerella cingulata* (in Colletotrichum state)	Probably universal; on other hosts in BSI, Cook Is., FP, NC, NH. Samoa, Wallis Is.
		Botryodiplodia theobromae	Probably universal; on other hosts in BSI, NC, NH. Samoa.
	Dieback and root rot	*Phytophthora cinnamomi*	Fiji.
Banana	Anthracnose	*Colletotrichum musae*	Probably universal; Fiji, NC, PNG.
	Black cross spot	*Phyllachora musicola*	Fiji, NH, PNG, Samoa.
	Bunchy top	Banana bunchy top virus	Fiji, Ellice Is., NC, ?NH, Samoa, Tonga, USTT, Wallis Is.
	Cordana leaf spot	*Cordana musae*	BSI, Fiji, FP, NC, NH, Norfolk Is., PNG, Samoa, USTT, Wallis Is.
	Finger rot	*Botryodiplodia theobromae, Colletotrichum musae*	Probably universal; Fiji.
	Freckle	*Macrophoma musae*	Fiji, NC.
	Malayan leaf spot	*Haplobasidion musae*	Probably widespread; Fiji, Samoa, Tonga.
	Mosaic	Cucumber mosaic virus	Probably widespread; Fiji, Samoa, Tonga.
	Panama disease	*Fusarium oxysporum* f. sp. *cubense*	?Fiji
	Rust	*Uredo musae*	Fiji, PNG, Samoa.
	Sigatoka disease	*Mycosphaerella musicola*	Probably widespread; Fiji, FP, NC, NH, Samoa, Tonga, USTT.
	Black leaf streak	*Mycosphaerella* sp.	Probably widespread; Fiji, FP, NC, PNG, Samoa, Tonga, USTT.
	Speckle	*Chloridium musae*	Probably widespread; Fiji.

APPENDIX

Host	*Disease*	*Pathogen*	*Distribution*
Batiki blue-grass (*Ischaemum indicum*)	Eye spot	*Curvularia* sp.	Fiji.
Bele *(Hibiscus manihot)*	Root and collar rot	*Phytophthora nicotianae* var. *parasitica*	Fiji.
Cabbage	Black rot	*Xanthomonas campestris*	Fiji, NC, NH, Norfolk Is., PNG, Tonga.
Carrot	Alternaria blight	*Alternaria dauci*	Fiji, NC, PNG.
	Root knot	*Meloidogyne incognita*	Fiji.
Centro *(Centrosema pubescens)*	Leaf spot	*Cercospora bradburyae*	BSI, Fiji, PNG.
Chilli	Mosaic	Cucumber mosaic virus?	Fiji, NC.
Chinese cabbage	Leaf rot	*Corticium sasakii*	Fiji.
	Mosaic and yellows	Cucumber mosaic virus?	Fiji.
	Soft rot	*Erwinia* spp.	Fiji.
	White rust	*Albugo candida*	Fiji, NC; on other hosts in Fiji, NC, PNG.
Citrus	Anthracnose	*Glomerella cingulata* (in Colletotrichum state)	BSI, Fiji, FP, NC, NH, PNG, Wallis Is.
	Blue and green mould	*Penicillium* spp.	Probably universal; Fiji.
	Canker	*Xanthomonas citri*	Fiji, PNG, USTT.
	Felt fungus	*Septobasidium* spp. (on scale insects)	Fiji, FP, NC, PNG.
	Foot rot, gum-mosis, and brown rot	*Phytophthora* spp.	Fiji, USTT.
	Greasy melanose	*Mycosphaerella horii*	Fiji.
	Heart rot	*Ganoderma* sp.	Fiji.
	Quick decline	Citrus tristeza virus	Probably widespread; Fiji, FP, NC.
	Melanose	*Diaporthe citri*	Fiji.
	Scab	*Elsinoe fawcettii*	BSI, Cook Is., Fiji, Guam, NC, NH, PNG, Samoa.
	Sooty mould	*Capnodium citri* (associated with insects)	Probably universal; Fiji.
	Woody gall	Citrus vein enation virus	Fiji.
Cocoa	Algal disease	*Cephaleuros* spp.	Probably universal on many hosts; Fiji, FP, NH; on coffee also in NC.
	Anthracnose	*Glomerella cingulata* (in Colletotrichum state)	Fiji.
	Black pod and canker	*Phytophthora palmivora*	Fiji, NH, PNG, Samoa, USTT.

APPENDIX

Host	*Disease*	*Pathogen*	*Distribution*
	Brown pod and canker	*Botryodiplodia theobromae*	Probably universal on cocoa in BSI, Fiji, NH, PNG, Samoa, USTT.
	Collar rot	*Corticium* sp. *Phellinus lamaoensis* *Ustulina deusta*	Fiji, NH, Samoa; on other hosts in FP, NC.
	Thread blight	*Corticium* sp.	Fiji, NH.
Coconut	Bleeding disease	*Ceratocystis paradoxa*	Probably universal on coconut; Fiji, NC, NH.
	Bud rot	*Phytophthora palmivora*	Fiji, FP, NC, NH, PNG.
	Grey leaf spot and blight	*Pestalotiopsis palmarum*	Probably universal on coconut; BSI, FP, NC, NH, PNG, Samoa, USTT, Wallis Is.
	Seedling blight	*Helminthosporium incurvatum* and other fungi	Fiji, PNG.
	Thread blight	*Corticium penicillatum*	Fiji, NH, PNG.
Coffee	Brown eye spot and berry blotch	*Cercospora coffeicola*	Fiji, FP, NC, PNG, Samoa.
	Root rot and dieback	*Phellinus lamaoensis*	Fiji, FP, NC, Samoa.
	Rust	*Hemileia vastatrix*	Fiji, NC, NH, PNG, Samoa.
	Thread blight	*Corticium koleroga*	Fiji, NC, PNG, Samoa.
Cucurbits	Downy mildew	*Pseudoperonospora cubensis*	Fiji, FP, Guam, PNG, Tonga.
	Gummy stem blight	*Mycosphaerella melonis*	BSI, Fiji, Pitcairn Is.
	Leaf spot	*Cercospora citrullina*	Fiji, NH, PNG, Tonga.
	Mosaic	Cucumber mosaic virus, Watermelon mosaic virus	Fiji.
	Powdery mildew	*Erysiphe cichoracearum* (in Oidium state)	Fiji, NC, Tonga.
Dalo (Taro)	Brown leaf spot	*Cladosporium colocasiae*	Fiji, NC, NH, PNG.
	Corm rot	*Phytophthora* spp.	BSI, Fiji.
	Shothole	*Phyllosticta colocasiicola*	Fiji
	Veinbanding	Unknown	Fiji
Duruka *(Saccharum edule)*	Fiji disease	Sugarcane Fiji disease virus	On sugarcane in BSI, Fiji, NC, NH, PNG, Samoa.
	Leaf scald	*Xanthomonas albilineans*	Fiji.
	Red rot	*Glomerella tucumanensis* (in Physalospora state)	On sugarcane in BSI, Fiji, Guam, NC, NH, Samoa, Tonga.
	Ring spot	Unknown	Probably universal
	Rust	*Puccinia kuehnii*	On sugarcane in Fiji, Guam, NC, PNG, Samoa.

APPENDIX

Host	*Disease*	*Pathogen*	*Distribution*
	Yellow spot	*Cercospora koepkii*	On sugarcane in BSI, Fiji, NC, PNG, Samoa.
Eggplant	Anthracnose	*Glomerella cingulata* (in Colletotrichum state)	Probably universal; Fiji.
	Leaf spot	*Cercospora egenula*	Fiji, NC, Tonga.
Eggplant	Leaf spot	*Corynespora cassiicola*	Fiji.
Ginger	Root knot	*Meloidogyne incognita*	Fiji.
	Soft rot	*Pythium gracile*	Fiji.
Haricot bean	Angular leaf spot	*Phaeoisariopsis griseola*	Fiji, NC, PNG.
	Halo blight	*Xanthomonas phaseolicola*	Fiji.
	Mosaic	Bean mosaic virus, Cucumber mosaic virus	Fiji.
	Pod rot	*Rhizoctonia* sp.	Fiji.
	Root and stem rot	*Fusarium* sp., *Glomerella* sp.	Fiji.
Jackfruit	Immature fruit rot	*Rhizopus artocarpi*	Fiji, NC.
Kumala (Sweet potato)	Leaf spot	*Cercospora batatae*	Fiji.
	Scab	*Elsinoe batatas*	BSI, Fiji, Guam, Niue, NC, NH, PNG, Tonga.
	Tuber rot	*Geotrichum* sp., *Erwinia* sp.	Fiji
Maize	Downy mildew	*Sclerospora sacchari*	Fiji.
	Leaf spot	*Cochliobolus heterostrophus* (in Helminthosporium state)	BSI, Fiji, NC, NH.
	Mosaic	Cucumber mosaic virus, Sugarcane mosaic virus	Fiji.
	Rust	*Puccinia polysora*	BSI, Fiji, NC, NH, PNG, Samoa, Tonga.
		Puccinia sorghi	Fiji, NC (probably intermixed with *P. polysora*).
	Sheath and pink ear rot	*Gibberella fujikuroi*	Fiji, Guam, NC, Samoa.
		Gibberella zeae	Fiji, NC, PNG.
Mango	Anthracnose	*Glomerella cingulata* (in Colletotrichum state)	Fiji, NH, PNG, USTT.
Onion	Neck rot	*Glomerella* sp., *Aspergillus flavus*, *Aspergillus niger*	Probably universal; Fiji.
Pangola grass	Rosette	Pangola rosette virus	Fiji.
Passion fruit	Anthracnose	*Glomerella cingulata* (in Colletotrichum state)	Probably universal on other hosts; Fiji.
	Brown spot	*Alternaria passiflorae*	Fiji, PNG.
	Collar rot	*Phytophthora* spp.	Fiji.

APPENDIX

Host	*Disease*	*Pathogen*	*Distribution*
	Mosaic and woodiness	Cucumber mosaic virus, Woodiness Virus.	Fiji.
	Thread blight	*Corticium solani*	Fiji.
Papaw	Fruit and root rot	*Phytophthora nicotianae* var. *parasitica*	Fiji, ?NC.
	Ripe fruit rot	*Botryodiplodia theobromae, Fusarium* spp., *Glomerella* spp.	Probably universal; Fiji.
	Shothole	*Corynespora cassiicola*	BSI, Fiji, NC, NH, PNG, Samoa.
Peanut	Bacterial wilt	*Pseudomonas solanacearum*	Fiji.
	Crown rot	*Aspergillus niger*	Fiji, PNG.
	Diplodia blight	*Botryodiplodia theobromae*	Fiji.
	Leaf rot	*Corticium sasakii*	Fiji.
	Leaf spot	*Mycosphaerella arachidicola*	Fiji, NC, PNG.
		Mycosphaerella berkeleyi	BSI, Fiji, NC, PNG, Tonga.
Pineapple	Base rot	*Ceratocystis paradoxa*	Probably universal on pineapple; Fiji, NC.
	Fruitlet core rot	*Cladosporium* sp., *Curvularia* sp., *Fusarium* sp., *Penicillium* sp.	Probably universal; Fiji.
	Heart and root rot	*Phytophthora spp.*	Fiji.
Potato	Bacterial wilt and brown rot	*Pseudomonas solanacearum*	Fiji, Tonga.
	Dry rot and seed-piece rot	*Fusarium* sp.	Probably universal; Fiji.
	Early blight	*Alternaria solani*	Fiji, NC (also on tomato).
	Scab	*Streptomyces scabies*	Fiji, NC, PNG.
	Soft rot	*Erwinia* spp.	Probably universal; Fiji.
	Wet rot	*Ceratocystis paradoxa*	Fiji.
Pulses (Cowpea, long bean, mung, urd)	Leaf spot	*Cercospora canescens*	BSI, Fiji, NC, NH, PNG, Tonga.
		Cercospora cruenta	BSI, Fiji, NC.
	Mosaic	Cucumber mosaic virus, Cowpea mosaic virus	Fiji.
	Powdery mildew	*Erysiphe polygoni* (in Oidium state)	Fiji.
	Rust	*Uromyces appendiculatus*	Fiji, NC, Norfolk Is., PNG.
		Uromyces vignae	BSI, Fiji, PNG.
Rice	Blast	*Pyricularia oryzae*	Fiji.
	Brown spot	*Cochliobolus miyabeanus* (in Helminthosporium state)	BSI, Fiji, NC, PNG.
	False smut	*Ustilaginoidea virens*	Fiji, PNG.
	Fool's rice	*Gibberella fujikuroi*	Fiji, NC, PNG.

APPENDIX

Host	*Disease*	*Pathogen*	*Distribution*
	(Bakanae disease)	(in Fusarium state)	
	Foot rot	*Ophiobolus oryzinus*	Fiji, PNG.
	Grey leaf spot	*Pyrenochaete oryzae*	BSI, Fiji.
	Narrow brown leaf spot	*Cercospora oryzae*	Fiji, PNG.
	Sheath rot	*Corticium sasakii*	Fiji, PNG.
	Stem rot	*Leptosphaeria salvinii*	Fiji, NC.
Sesame	Leaf spot	*Cercospora sesami*	Fiji.
Sorghum	Head blight	*Fusarium* spp., *Glomerella* sp., *Nigrospora* sp., *Phyllosticta* sp.	Fiji.
	Leaf spot	*Helminthosporium turcicum*	Fiji.
	Rust	*Puccinia purpurea*	Fiji, NC, PNG.
	Top rot	*Gibberella fujikuroi*	Fiji.
Tapioca	Leaf spot	*Cercospora henningsii*	Probably universal; BSI, Fiji, FP, NC, NH, Samoa, Tonga, Wallis Is.
Tea	Grey blight	*Glomerella* sp., *Pestalotia* sp.	Fiji.
	Red rust	*Cephaleuros pulvinatus*	Fiji.
	Root knot	*Meloidogyne incognita*	Fiji.
Tobacco	Frog eye	*Cercospora nicotianae*	Probably universal; BSI, Fiji, NC, PNG, Tonga, Wallis Is.
	Mosaic	Tobacco mosaic virus	Fiji.
	Root knot	*Meloidogyne incognita*	Fiji.
Tomato	Anthracnose	*Glomerella phomoides* (in Colletotrichum state)	Fiji, NC.
	Bacterial spot	*Xanthomonas vesicatoria*	Fiji.
	Bacterial wilt	*Pseudomonas solanacearum*	Fiji, PNG, USTT.
	Buckeye rot	*Phytophthora nicotianae* var. *parasitica*	Samoa
	Corynespora leaf spot	*Corynespora cassiicola*	BSI, Fiji.
Tomato	Leaf mould	*Cladosporium fulvum*	Fiji, NC, PNG, Samoa.
	Mosaic	Tobacco mosaic virus	Fiji.
	Root knot	*Meloidogyne incognita*	Fiji, PNG.
	Shoestring	Cucumber mosaic virus	Fiji.
	Streak	Mixed viruses	Fiji.
Yaqona *(Piper methysticum)*	Wilt and stem rot	Unknown	Fiji.
Yam	Leaf spot	*Cercospora* sp.	Fiji.

APPENDIX

References:

1. Archives du laboratoire de phytopathologie du Centre ORSTOM de Nouméa.
2. Huguenin, B. Microfungi of New Caledonia. FAO Plant Protection Committee for Southeast Asia Technical Document No. 57 (Cyclostyled. Centre ORSTOM de Nouméa). 1967.
3. Morwood, R.B. A preliminary list of plant diseases in Fiji. Fiji Agr. Journal 27 : 51-54. 1956.
4. Review of Applied Mycology Vols. 1-46. 1922-1967. Commonwealth Mycological Institute, Kew, Surrey, England. (Records compiled by Mr. Anthony Johnston, Assistant Director, CMI).
5. Shaw, D.E. Plant pathogens and other micro-organisms in Papua and New Guinea. Dept. Agr., Stock & Fisheries Res. Bulletin No. 1. 1963.

Dd. 151944 K 9/70 10